HUMAN EMBRYONIC
STEM
CELLS

SECOND EDITION

CABRINI COLLEGE LIBRARY
610 KING OF PRUSSIA ROAD
RADNOR, PA 19087

ANN A. KIESSLING
HARVARD MEDICAL SCHOOL

SCOTT C. ANDERSON

JONES AND BARTLETT PUBLISHERS

Sudbury, Massachusetts

BOSTON   TORONTO   LONDON   SINGAPORE

#656 17482

World Headquarters

Jones and Bartlett Publishers
40 Tall Pine Drive
Sudbury, MA 01776
978-443-5000
info@jbpub.com
www.jbpub.com

Jones and Bartlett Publishers Canada
6339 Ormindale Way
Mississauga, Ontario L5V 1J2
CANADA

Jones and Bartlett Publishers International
Barb House, Barb Mews
London W6 7PA
UK

Jones and Bartlett's books and products are available through most bookstores and online booksellers. To contact Jones and Bartlett Publishers directly, call 800-832-0034, fax 978-443-8000, or visit our website www.jbpub.com.

Substantial discounts on bulk quantities of Jones and Bartlett's publications are available to corporations, professional associations, and other qualified organizations. For details and specific discount information, contact the special sales department at Jones and Bartlett via the above contact information or send an email to specialsales@jbpub.com.

Copyright © 2007 by Jones and Bartlett Publishers, Inc.

ISBN-10: 0-7637-4386-0
ISBN-13: 978-0-7637-4386-4

All rights reserved. No part of the material protected by this copyright may be reproduced or utilized in any form, electronic or mechanical, including photocopying, recording, or by any information storage and retrieval system, without written permission from the copyright owner.

Production Credits
Chief Executive Officer: Clayton Jones
Chief Operating Officer: Don W. Jones, Jr.
President, Higher Education and Professional Publishing: Robert W. Holland, Jr.
V.P., Design and Production: Anne Spencer
V.P., Sales and Marketing: William Kane
V.P., Manufacturing and Inventory Control: Therese Connell
Acquisitions Editor, Science: Cathleen Sether
Managing Editor, Science: Dean W. DeChambeau
Editorial Assitant: Molly Steinbach
Senior Production Editor: Louis C. Bruno, Jr.
Marketing Manager: Andrea DeFronzo
Text and Cover Design: Anne Spencer
Illustrations: Elizabeth Morales
Composition: International Typesetting and Composition
Printing and Binding: Malloy
Cover Printing: Lehigh Press
Cover Photo: Courtesy of Bedford Research Foundation. Photographed by Ann A. Kiessling.

About the cover: A photograph of a human embryonic stem cell colony stained with Hoechst dye to visualize DNA.

Library of Congress Cataloging-in-Publication Data
Kiessling, Ann A.
    Human embryonic stem cells / Ann Kiessling, Scott Anderson.—2nd ed.
       p. ; cm.
    Includes bibliographical references and index.
    ISBN-13: 978-0-7637-4386-4 (alk. paper)
    1. Embryonic stem cells. I. Anderson, Scott, 1951—II. Title.
    [DNLM: 1. Stem Cells. 2. Embryo—cytology. QU 325 K47h 2006]
    QP277.K54 2006
    612.6'4—dc22   2006010791

Printed in the United States of America
10 09 08 07 06   10 9 8 7 6 5 4 3 2 1

To Mike and Larry, men with hemophilia; wonderful sons, husbands, and fathers who were infected with the AIDS virus and hepatitis C by tainted blood clotting factors prescribed to treat their hemophilia; men who died young, before stem cell therapy could cure them.

And to Mom and Dad, whose daily, noble struggles with Parkinson's disease and heart failure contributed urgency to the writing of this text.

Brief Contents

Contents

Preface

Advances in stem cell science have raised broad ethical concerns. Remarkably, in the three and a half years since the first edition of *Human Embryonic Stem Cells*, the nature and polarity of the ethical concerns have not changed.

That the debate is taking place speaks highly of our society's genuine concern for the protection of its most vulnerable members. What is missing from the debate process, however, is a thorough understanding of the biology involved in stem cell research. The debaters opposed to research that would destroy human embryos liken the process to abortion and murder. Those in favor of the research justify the destruction of embryos for the sake of suffering persons. Neither viewpoint is accurate. Importantly, many of the facts necessary to establish an entirely accurate viewpoint are not now known because the necessary research has not been done.

The emerging field of human embryonic stem cell biomedicine crosses many disciplinary boundaries, including those of reproductive biology, embryology, cell biology, molecular biology, endocrinology, immunology, fetal medicine, transplantation medicine, and surgery. *Human Embryonic Stem Cells* was written to meet the needs of not only science students with a basic understanding of the principles of cell biology but also science writers, politicians, teachers, medical students, physicians, nurses, veterinarians, and biomedical scientists who may not be familiar with the various stem cell disciplines.

The term *embryonic* has been used inaccurately by both scientists and the public, thus, deepening the confusion on both sides of the debate. It is used as a substitute for the more accurate term *pluripotent*, which describes a cell with the capacity to give rise to cells of all tissue types in the human body. Clearly what is needed is a single reference that provides the basic information from these multiple disciplines as it pertains to the science of human pluripotent stem cells, one type of which is embryonic stem cells. Because the field is so new, and the social debate so heated, it would be especially valuable if such a reference could be understood by the public as well as by the scientists in the field. This is the goal of this undertaking.

Because of the newness of stem cell science, we have included some of the history of scientific endeavors that advanced the field. These are presented throughout the text as sidebars. Not only do they enhance the understanding of how the field has evolved, but they provide scientific details important to understanding the biology as we now know it. For the sake of scientists, some chapters contain more advanced descriptions of what is known about some cell processes. This information is not essential to understanding what follows in succeeding chapters. The interested person can skip the cellular details without losing the thread of the overall explanation. Scientific facts that are still missing and are in need of discovery are emphasized in each.

We apologize to our colleagues whose work is not featured as they feel appropriate. We anticipate hearing from them and including their suggestions in future editions. We thank our families and Steven Hagan for patience and support during this project. We also thank Rachel Francis for bringing a recent veterinary school graduate's view to the information presented in the first edition, a task she completed while tending to AK's first grandchild, Autumn Rose. The Jones and Bartlett team, especially Lou Bruno, Cathleen Sether, Steve Weaver, and Anne Spencer, were as competent and supportive of this endeavor as with the first edition; they are highly skilled at managing the talent and idiosyncrasies of their science authors. Thanks go to artist Elizabeth Morales for so skillfully rendering the ideas that were often presented to her as incomplete dialogue.

We have written this book with the view that there is no substitute for knowledge when attempting to formulate an opinion about a new and awesome biomedical development. Those who are firmly on one side or the other of the debate may not be swayed from their positions by the information presented. However we hope that those scientists, developing scientists, teachers, writers, politicians, physicians, and patients struggling to fully understand the field, its therapeutic potential, and its potential pitfalls will find that by reading the book, they come to a new understanding of the field of human pluripotent stem cells and will be able to participate fully in its development and its oversight. Given the broad scope of therapeutic potential, there is great likelihood that all readers of this text will benefit from stem cell technologies at some point in their lives.

A. Kiessling
S. Anderson

Foreword

Welcome to the future! *Human Embryonic Stem Cells* is a special book that boldly addresses a subject that is more controversial and therapeutically promising than the first kidney transplant in 1954. Kidney transplants are now routine and save many lives. The same potential exists for embryonic stem cells. Ann Kiessling and Scott Anderson address the burgeoning field in this book, which makes stem cell science understandable to the general public, legislators, college students, and graduate students. It's a great read.

Three years ago, in 2003, I was inspired by the publication of the first edition of *Human Embryonic Stem Cells* to initiate a course called Embryonic Stem Cells at Northeastern University. As a reproductive immunologist, who has conducted research for 36 years on genes that mediate the survival of preimplantation embryos, I was ready to embrace the new field of embryonic stem cell research. Preimplantation embryos can develop into babies or embryonic stem cells. It is the latter that has the potential to differentiate into any tissue in the body with the expectation that extensive research on embryonic stem cells will result in cures for a myriad of diseases, including diabetes and Parkinson's disease, and for spinal cord injury. I ran my Embryonic Stem Cell course for seniors and graduate students. I used *Human Embryonic Stem Cells* as the text for the course and supplemented the chapter information with papers from the original literature. The students gave PowerPoint presentations on the chapters and original literature, followed by extensive discussions of the material in class. Overall the course has been an outstanding success. The following are some comments from the students on the textbook:

- "The way Kiessling and Anderson present information on human embryonic stem cells is very well organized."
- "This book is a great start to educating people on a very new, controversial topic."
- ". . . the writing style used by Kiessling and Anderson was clear and concise and allows for easy understanding of the concepts to all audiences."
- "One of the strong points of the text is how it deals with the ethical debate central to stem cells gracefully, but without shying away from important issues because they are controversial."

This second edition of *Human Embryonic Stem Cells* contains up-to-date information on this fast-moving field. All of the successful attributes of the first edition have been retained including the excellent diagrams, tables, sidebars, and glossary. The Additional Readings at the end of each chapter have been expanded and updated. In addition, a new section of Review Questions has been added to end of each chapter, which should inspire further thought about the material presented. Many biochemical, immunological, and genetic techniques are mentioned throughout the book, and explanations of these techniques are now available on the publisher's website (www.bioscience.jbpub.com). These additions will be most welcome the next time I teach my Embryonic Stem Cell course.

Aside from serving as a textbook, this book can also be a primer for researchers entering this new field, for legislators debating the legal and ethical issues attached to human embryonic stem cell research, and for interested public citizens. Dr. Kiessling is to be commended for her work with the Massachusetts state legislature, which legalized human embryonic stem cell research in Massachusetts in 2005, and for her work on the Research Standards Working Group of the California Institute for Regenerative Medicine, and Connecticut's Stem Cell Advisory Committee. The next step is to raise money for this endeavor. I challenge the readers of this second edition of *Human Embryonic Stem Cells* to work with their state and national legislators to secure funding at the state and national levels for one of the most promising fields in all of biological research.

Carol Warner
Matthews Distinguished University Professor
Department of Biology
Northeastern University
Boston, MA

PART

I

The Basics

CHAPTER 1

The Nature of Stem Cells

*All truth passes through three stages. First, it is ridiculed. Second,
it is violently opposed. Third, it is accepted as being self-evident.*

Arthur Schopenhauer, 1788–1860

OVERVIEW

Advances in medicine follow advances in biology. Most biomedical research improves our understanding of the details of known biologic processes. For example, once it was known that cancer could result from mutations in single genes in single cells, discoveries of mutated genes in cancer cells abounded, and dozens have now been identified. Fundamentally, new biologic discoveries occur rarely, however, and are dependent on new ways of thinking and new technologies. A remarkable number of fundamentally important processes, such as embryonic development and memory, are poorly understood and await new insights. The lack of understanding of embryonic development is especially puzzling given its key importance to every living thing. Those processes that ensure reproduction are the most important to survival of the species and, hence, may be the most robust and redundant.

Most biomedical research focuses on adult diseases, which can be broadly divided into three problems at the cellular level: too much cell division, too little cell division, and defective cell function. Too much cell division results in tumors, which are sometimes cancerous. Too little cell division results in the inability to repair damaged tissues and organs, such as severed spinal cords and weak heart muscle. Defective cell function results in diseases such as hemophilia (lack of synthesis of blood clotting factors by liver cells) and Parkinson disease (lack of dopamine production by brain cells). Historically, treatments for adult diseases have been surgery or pharmacologic intervention. For example, major research efforts have developed the methods for heart transplantation to "cure" weak hearts. Dozens of drugs have been developed to correct abnormal heart rhythms and support heart muscle weakened by high blood pressure or heart attack. The concept of repairing weak heart muscle cells by encouraging the growth of new muscle cells has, until recently, not been considered possible. This new way of thinking about treating disease has brought stem cells to the fore-front. The term **stem cells** is used to describe those cells that serve as a normal reservoir for new cells needed to replace damaged or dying cells. As described in the following chapters, a fundamental characteristic of stem cells is the lasting ability to multiply when called on. Stem cells are broadly divided into three groups: adult stem cells, fetal stem cells, and "embryonic" stem cells (Table 1.1).

Stem cell: a cell capable of unlimited cell cycles with the capacity to cease being a stem cell and become one of a wide variety of specialized cells that compose a specific tissue or organ.

TABLE

1.1 Classes of Stem Cells

Source	Description	Examples
	ADULT STEM CELLS	
Adult organs	Reservoir of cells partially committed to type of tissue in organ	Liver, bones, bone marrow, lining of gut, spermatozoa, olfactory neurons, and skin
	FETAL STEM CELLS	
Fetal tissues	Precursor cells more abundant in growing organs; some uncommitted cells	Heart muscle cells, brain cells, pancreatic cells, germ cell precursors to sperm and eggs
	"EMBRYONIC" STEM CELLS	
Embryos	Derived from inner cell mass cells of blastocysts created by fertilization of eggs with sperm	Frozen embryos left over from in vitro fertilization procedures
Parthenotes*	Derived from inner cell mass cells of blastocysts created by chemical/electrical stimulation of unfertilized eggs	Monkey parthenote stem cells are a valuable research model; no human parthenote stem cells
Ovasomes	Derived from blastocysts created after complete exchange of genetic information by nuclear transplantation	Cell lines available from mouse and cow; process related to animal cloning

*The photo to the right, which appears on all chapter opening spreads, is that of a human parthenote morula taken in 2005. Courtesy of A. A. Kiessling.

Adult Stem Cells

After their formation in the fetus, some tissues and organs continue to maintain a population of stem cells throughout childhood and into adulthood. Even though these cells may appear early in development, they are called adult stem cells. Examples are skin cells and blood vessels that readily repair damage and can actually regenerate. Another example is spermatogenesis. The fetal testes are populated with stem cells that are not actually stimulated to produce more cells until puberty, but once sperm production is initiated, it continues throughout the life of the male. Billions of new sperm are produced daily. Another example is bone marrow, which maintains a population of blood cell stem cells that also produce billions of new blood cells each day. Cells that line the stomach and intestines are also regularly replaced. These examples illustrate fundamental features of adult stem cells: They maintain the ability to divide throughout life and give rise to specific cell types. Blood vessel stem cells give rise to blood vessels but cannot give rise to sperm, and spermatogonia cannot give rise to blood. Hence, the developmental potential of adult stem cells is restricted.

This restriction in developmental potential of adult stem cells is the essence of the need for new sources of stem cells. Those tissues and organs that do not maintain a population of stem cells throughout life cannot repair themselves. Why some tissues maintain stem cells and others do not is not understood and is an area of active research described in Chapter 11. Tissues deficient in stem cells include brain, heart, spinal cord, eye, and kidney. There have been recent reports that a small population of stem cells also may be present in these tissues, and the challenge to scientists is to isolate them in the laboratory and encourage their multiplication into sufficient numbers to be therapeutically useful. One hope is to accomplish this from a biopsy of the failing organ. Another hope is to isolate stem cells from adult organs donated by individuals before their death for such a purpose. The enormous human and economic value of creating heart muscle stem cells or kidney stem cells to tissue match to all persons with failing hearts or kidneys is clear and an active area of research.

Proof of principle of the effectiveness of this approach is exemplified by bone marrow transplantation, which is, in fact, ongoing therapeutic use of adult stem cells. Dr. E. Donnell Thomas shared the Nobel Prize in Medicine in 1990 for his pioneering work in bone marrow transplantation, which began 50 years ago (see sidebar). Historically thought of as tissue transplantation, considerations of the role of bone marrow in blood formation reveal marrow transplantation is essentially bone marrow stem cell therapy.

Blood functions to carry oxygen, nutrients, hormones, and immune responsive cells to all parts of the body. Carrying out these functions requires continual cell death and replacement for several reasons. Bleeding or donating a unit of blood, for example, stimulates blood cell stem cells to multiply in the marrow to replenish the lost blood cells. Another example is replacement of the blood cells

Bone Marrow Transplantation
E. Donnell Thomas

In 1990, E. Donnell Thomas, M.D., shared the Nobel Prize in Physiology or Medicine with Joseph E. Murray, M.D. Both men pioneered medical therapies involving transplantation of tissues from one person to another. Dr. Murray pioneered kidney transplantation as a treatment for kidney failure, and Dr. Thomas pioneered transplantation of bone marrow cells as a treatment for leukemia and aplastic anemia. A fundamental difference between kidney transplantation and bone marrow transplantation is that the transplanted kidney is fully formed, but bone marrow cells must continue to proliferate and give rise to peripheral blood cells throughout the life of the transplant recipient. Thus, bone marrow transplantation, initially explored over 60 years ago, was the first form of stem cell therapy.

During Dr. Thomas' student years at Harvard Medical School (1943–1946), he became interested in leukemia and bone marrow. Research on the health effects of exposure to radiation (primarily x-rays) revealed that death resulted from loss of cells in the bone marrow, which led to severe anemia and loss of immune function. One possibility that intrigued Dr. Thomas was that the irradiation destroyed "stimulating factors" that promoted growth of bone marrow cells, and he spent a year at the Massachusetts Institute of Technology studying factors that stimulated yeast cell division. In 1955, work by Main and Prehn demonstrated that a mouse treated after radiation exposure by transplanted bone marrow also did not reject skin grafted from the marrow donor. Thus, the immune system of the irradiated mouse had been reconstituted from the transferred bone marrow, providing strong evidence that bone marrow contained blood and immune system stem cells.

From 1955 to 1962, while an affiliate of Columbia University, Dr. Thomas established the basic concepts for successful bone marrow transplantation using dogs as a model system for humans. In 1963, he was recruited to the Department of Medicine at the University of Washington School of Medicine in Seattle, where he compiled a team of physician researchers who by the early 1970s had demonstrated that patients with leukemia, aplastic anemia, or genetic diseases could be cured by marrow transplantation. Such procedures have now become routine medical care, and thousands of successful and curative bone marrow transplantations are conducted annually worldwide. (www.nobel.se/laureates/medicine)

lost as a result of fighting an infection. The process of replacing the cells in peripheral blood actually involves several intermediate stages of cell differentiation after division of the stem cell (Figure 1.1). A fundamental difference between the bone marrow stem cell and the differentiating blood cells is the loss of the ability to divide again. There may be several stages of differentiation; some stages can divide again and others cannot. Cells not fully mature but fully committed to one type of differentiated cell are termed *precursor* or *progenitor cells* rather

Cell differentiation: the process of becoming a fully mature nondividing cell with specialized expression of gene products needed to carry out specific tissue functions.

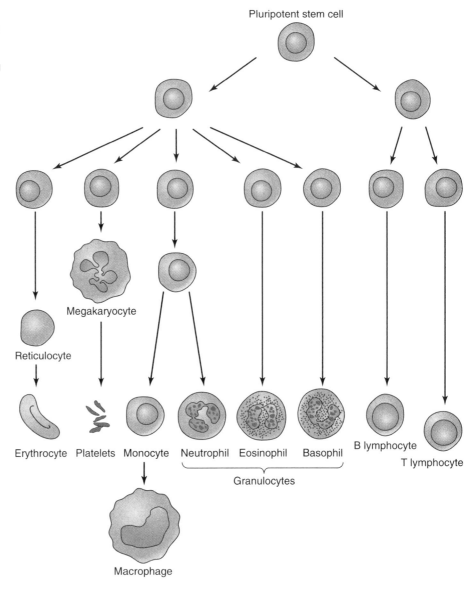

FIGURE 1.1
Blood cell pathways. All blood cells are thought to arise from a single multipotent blood stem cell. The stem cell divides to maintain a pool of multipotent stem cells that undergo further division, with some stem cells leaving the pool to enter pathways to differentiate into the many types of blood cells needed for daily life. An initial differentiation step gives rise to either lymphoid or myeloid "progenitor" cells, the intermediate cell type for the adult cells.

Pluripotent stem cell

Megakaryocyte

Reticulocyte

Erythrocyte   Platelets   Monocyte   Neutrophil   Eosinophil   Basophil   B lymphocyte

Granulocytes

T lymphocyte

Macrophage

than stem cells, but the distinction between stem cells and precursor/progenitor cells is not always clear.

As discussed in Chapter 11, bone marrow is no longer the only source of blood cell stem cells. Under certain conditions, venous blood can be enriched with bone marrow stem cells that can then be isolated by taking blood from a vein instead of by drilling a hole in a bone. This advance has not only improved our understanding of the equilibrium between bone marrow and peripheral blood, it has also markedly improved the availability of blood stem cells and made treatment far less painful.

Fetal Tissue Stem Cells

By approximately 3 months of pregnancy, organs are formed (described in Chapter 9). The developing human fetus is only a couple of inches long and weighs about 1 ounce, but primitive organs are in place. During the remaining 6 months of pregnancy, the organs enlarge and develop the ability to function independent of the placenta. Because the fetus is growing rapidly, all tissues and organs, including the brain, contain stem cells. It is for this reason that stem cell researchers are interested in studying fetal tissues. As described in Chapters 9, 11, 13, and 14, much progress has been made, but many questions remain.

Fetal tissue research has long sparked debate between scientists and some social groups. The fear seems to be that some pregnancies will be terminated (and perhaps created) just for research. The concern speaks to the high value our society places on individual human life and our interest in providing social defenses for those members of our society who cannot defend themselves. In many respects, this fear represents substantial progress in our sense of humanity and is the antithesis of the values of terrorists for whom individual life is less important than an ideologic goal.

The fear also relates to distrust of science and technology. There is an underlying belief that appropriate oversight of fetal tissue research cannot be achieved because scientists are too zealous in their pursuit of their research interests to be mindful of restraints. This underlying belief is also fueling the embryonic stem cell debate. Scientists must accept responsibility for the distrust of some elements of society. The distrust stems from many areas, including developments that are clearly unsafe, like nuclear weapons, and seemingly unnatural, like genetically engineered foods and assisted reproduction technologies. Profit motives, as exemplified by the pursuit of patents, add to the sense of distrust of scientists. Scientific language is replete with abbreviations for complex cell processes that make it difficult for nonscientists to follow the research, even though many of the concepts are easily understandable. The communication gap heightens the distrust. Bridging the communication gap between science and society is essential for complete acceptance of fetal and embryonic stem cell research.

It is also clear, however, that some social groups seeking to advance their own specific views attempt to misrepresent the goals of scientists. For example, the concept that fetal tissue and embryonic stem cell research cannot be conducted ethically is often advanced by groups who also seek to outlaw abortion. Some of those groups conduct large active lobbying programs in Congress to promote federal laws against all forms of embryonic stem cell research, claiming similarities to abortion. Fully informed decisions by ethicists, the general public, and legislators require understanding the biology of human embryonic development and stem cells.

Embryonic Stem Cells

Before organs are formed, embryos of all species comprise a collection of cells with potential to give rise to many different organs and tissues. Such cells with plural potentiality are termed *pluripotent stem cells*. First derived from mouse embryos, as described in Chapter 8, embryonic stem cells were found to have remarkable plasticity. They divide endlessly in laboratory culture dishes and maintain the ability to differentiate into numerous types of cells when exposed to the appropriate growth factors. For example, nerve growth factor, a protein hormone, can induce differentiation of stem cells in culture into mature nerves, called neurons, capable of transmitting electrical and chemical signals remarkably similar to nerves in the body (Figure 1.2). This holds great promise for repair of neurologic diseases such as Alzheimer and Parkinson and spinal cord injury. The problem is finding a source of pluripotent stem cells.

The proliferation of "leftover" human embryos as a result of aggressive hormonal treatment of women seeking to achieve pregnancy through assisted reproductive technologies (Table 1.2) has raised the possibility of deriving embryonic stem cells from human embryos by the same methods as applied to the mouse embryonic stem cells, which is described in Chapter 12. This area of research has been met with social resistance similar to fetal tissue research and has sparked additional controversy related to when, exactly, an embryo becomes human. The point at which human life is thought to begin varies among religions, so religious views play a substantial role in the debate.

From the standpoint of medical treatments, embryonic stem cells have the same tissue rejection concerns as stem cells derived from fetuses. Tissue compatibility problems may be less severe for fetal and embryonic stem cells than with adult organ transplants, but the cells still express the proteins encoded by the genes of the embryo when they undergo differentiation. Fortunately, new advances in the manipulation of cells and eggs, described in the following chapters, holds the promise of creating stem cells for each patient in need without creating a unique embryo. One problem, however, is that some of the new methods are the same as those used for cloning individuals. Social groups already opposed to fetal tissue research and/or the generation of embryonic stem cells from abandoned human embryos are truly aghast at the concept of pursuing technology that could also clone a human. Informed decisions for or against, however, require a thorough understanding of the basic biology.

FIGURE 1.2
Parthenote stem cells derived from the blastocyst of an activated monkey egg can differentiate into functioning neurons. (Modified from Cibelli, J. B., et al. [2002]. Parthenogenetic stem cells in nonhuman primates. Science 295: 819.)

Stem cells   At one week   At one month

TABLE 1.2	Embryos Created by Assisted Reproductive Technologies in the United States			
Year	Cycles of Hormone Treatments	Estimated Fertilized Eggs/Cycle	Average Number Transferred	Number of "Leftover" Embryos
1997	71,826	5 to 10	3 to 4	71,826 to 502,782
1999	86,822	5 to 10	3 to 4	86,822 to 607,754
2001	84,824	5 to 10	3	169,648 to 593,768
2003	95,723	5 to 10	2 to 3	191,446 to 765,784

Values are estimates because the Centers for Disease Control and Prevention does not report data relative to numbers of eggs harvested and fertilized; the exact number of cryopreserved embryos is unknown.

Parthenote Stem Cells

As discussed in Chapters 4 and 5, eggs may become activated without fertilization by sperm, a process known as parthenogenesis, from the Greek "virgin beginning." The activated egg undergoes early development to blastocysts in a manner similar to a fertilized egg, but it cannot give rise to offspring. That parthenote blastocysts may be a useful source of pluripotent stem cells has been demonstrated by the monkey parthenote cell line produced in 2002, which is being used as a model for stem cell therapy by laboratories all over the world.

Nuclear Transplant Stem Cells

As described in the following chapters, a fundamentally important biologic advance was the realization, in the early 1900s, that all human cells contain all chromosomes. In this way, each cell of each tissue and organ is a replicate reservoir of all the genes of each person. The single exception to this paradigm are mature sperm, which contain half the number of chromosomes of all other cells, leading to equal populations of X-bearing (female) and Y-bearing (male) sperm.

This led to the question of whether or not the chromosomes from adult cells could support the development of a completely new individual, theoretically identical to the adult. If true, then the cell differentiation needed for organ formation would be the result of cellular controls on the expression of individual genes. If false, then the cell differentiation needed for organ development could result from irreversible changes to the genes. There was also the possibility that the genes of some types of cells, but not all cells, were irreversibly altered. For example, it seemed possible that cells from those tissues with regenerative capacity, such as bone marrow stem cells or the lining of the stomach and intestines, might not contain irreversibly altered genes, whereas spinal cord nerves that do not regenerate may contain irreversibly altered genetic information.

One way to test these possibilities, as described in Chapters 7 and 10, was to transfer a nucleus into an egg whose chromosomes had been removed (**enucleated**) to determine whether the transferred nucleus could guide the development of an individual. Because this process involves manipulation of single cells, the egg and the cell donating the nucleus, it was termed *cloning*. The term stems from the ancient Greek "klon," which means twig, and refers to the fact that a twig could give rise to an identical tree. Ultimately, as described in Chapter 10, it was demonstrated that adult nuclei retain the potential for guiding the development of an entire animal, although this potential has not been demonstrated for cells from all adult tissues. The advantages of cloned animals to breeders, researchers, and pet owners are clear and even compelling, but the real power of the technology lies in the creation of stem cells for research and treatment of diseases.

Such nuclear transplant stem cells would be genetically identical to the adult and capable of giving rise to replacement cells for all body tissues. Gene replacement techniques could correct genetic defects in the stem cells to enable them to correct inherited disorders such as hemophilia and some forms of diabetes. The benefit to sufferers of now incurable diseases is clear.

The challenge before us is to provide the understanding among opposing social groups, religious interests, and legislators to allow the biology to move forward with societal controls that will not deny these powerful treatment options to the millions suffering from incurable diseases.

Enucleated: the process of removing cellular genetic information either in the form of an interphase nucleus or as chromosomes.

Human Cloning

Concerns about cloning a human range from religious views that it is unholy to fear of eugenics and the creation—in vast "clone farms"—of genetically selected individuals, generally referred to as some form of superbeing. The arguments and fears surrounding cloning serve to force society to examine what it truly values. The strong opposition at this time to even developing the technology speaks for the high value we place on human life and its attendant dignity. However, the opposition also raises significant inconsistencies in current social practices, such as the death penalty and the tolerance of polygamy. Is it mere coincidence that Osama Bin Laden, the terrorist who seeks to impose his way of life on the entire world, is 1 of 50 or so children from a polygamous father clearly seeking to enrich humanity with his own genetics? From the Old Testament's Abraham to modern day Josiah Smith, man has manifested a passion to pass on his genes. It needs to be recognized that, to some individuals, cloning may represent simply a more advanced way to achieve this, not a new desire.

There are many reasons besides social and religious concerns not to clone a human. The experience with animals has shown that cloned individuals have a high incidence of failure to thrive. Because this is not true of identical twins, who are clones of each other, there is a problem with the current technology. Given this knowledge, it is at the least irresponsible to attempt to clone a human until the problems are solved in animal models. However, given the notoriety it will bring, it is likely to be attempted within the foreseeable future. Society then will be forced to face a new set of fears. On the positive side, the technology could be modified to allow reproduction by men who have no sperm and women who have no eggs. This possibility could be viewed as providing a social and genetic balance against those highly fertile individuals clearly bent on proliferating their own gene pool.

Summary

Diseases can be broadly divided into three problems at the cellular level: too much cell division, too little cell division, and defective cell function. Diseases resulting from too little cell division and defection cell function are the most immediate targets for stem cell therapy.

There are several types of stem cells. Stem cells derived from the inner cell mass of blastocysts have the capacity to give rise to all the cell types in all adult tissues; for this reason they are termed *pluripotent*. Embryonic blastocysts arise from fertilized eggs, parthenote blastocysts arise from eggs activated without being fertilized by sperm, and nuclear transplant blastocysts arise from eggs whose chromosomes have been replaced with those of another cell. Stem cells cultured from the inner cell masses of each type of blastocyst are termed *embryonic stem cells, parthenote stem cells, and nuclear transplant stem cells* (also termed *ovasomes*), respectively.

Some adult tissues retain a reservoir of tissue-specific stem cells with the capacity to replace defective cells as needed. Such cells are termed *adult stem cells*. Bone marrow stem cells are an example of adult stem cells. For the most part, adult stem cells are restricted to regenerating cells of their tissue of origin, but studies to determine whether their regenerative potential can be broadened to other tissue types are ongoing in many laboratories.

Tissues in a developing fetus contain cells with regenerative capacity inter-mediate between pluripotent stem cells and adult stem cells. Fetal tissue stem cells have provided excellent model systems for studies of the regenerative capacity of stem cells, but they are a limited resource and not considered an answer to ther-apeutic needs.

Review Questions (True or False)

1. Pluripotent stem cells can give rise to all types of adult tissues.
2. Blastocysts with inner cell masses composed of pluripotent stem cells can form from fertilized eggs, from eggs activated without fertilization, and from eggs whose chromosomes have been replaced by those from another cell.

3. Stem cell therapy began with bone marrow transplantation several decades ago.
4. Adult stem cells are not considered therapeutically valuable.
5. The process of developing mature blood cells involves many intermediate stages of cell differentiation, some of which also undergo cell division.

ADDITIONAL READINGS

ART Reports, www.cdc.gov/ART

Cibelli, J. B., Grant, K. A., Chapman, K. B., Cunniff, K, Worst, T, Green, H. L., Walker, S. J., Gutin, P. H., Vilner, L, Tabar, V., Dominko, T., Kane, J, Wettstein, P. J., Lanza, R. P. Studer, L, Vrana, K. E., and West, M. D. (2002). Parthenogenetic stem cells in nonhuman primates. *Science* 295: 819.

Committee on the Biological and Biomedical Applications of Stem Cell Research, Board on Life Sciences (2002). *Stem Cells and the Future of Regenerative Medicine.* Washington, DC: National Academies Press.

Green, R. A. (2001). *The Human Embryo Research Debates.* New York: Oxford University Press.

Kiessling, A. A. (2004). *What is an embryo? Connecticut Law Rev* 36: 1051–1092.

Kirschstein, R., and Skirboli, L. (2001). *Stem Cells: Scientific Progress and Future Research Directions.* Bethesda, MD: National Institute of Health. www.nih.gov

Quinn, K.P. (2004). The politics of embryonic discourse. *Connecticut Law Rev* 36: 1163–1169.

CHAPTER 2

The Dividing Cell

The uniformity of earth's life, more astonishing than its diversity, is accountable by the high probability that we derived, originally, from some single cell, fertilized in a bolt of lightning as the earth cooled.

Lewis Thomas, The Lives of a Cell

OVERVIEW

Robert Hooke coined the term *cell* from observations of cork made with a simple microscope in 1665. A few years later, Antony van Leeuwenhoek described sperm, red blood cells, and some bacteria. However, it was not until the 1830s that the microscopy studies of Theodor Schwann and Matthias Schleiden reached the same conclusion for animals and plants, respectively: All organisms are composed of individual cells. Much has been learned in the ensuing 170 years, but much remains to be discovered.

Every advance in microscopy was accompanied by new discoveries about cells. Their most obvious features (Figure 2.1) are their outer (plasma) membrane, their nucleus, also enclosed with a (nuclear) membrane, and the fact that cells only arise from other cells. In this regard, all human life is a continuum. Further advances in microscopy also revealed other structures in the cell cytoplasm, including mitochondria, which generate energy, and ribosomes, which organize protein synthesis, as well as the Golgi apparatus, which is responsible for processing newly synthesized proteins and other large molecules synthesized by the cell. Importantly, each of these structures must also be replicated for division into the daughter cells.

Tissues and organs are made up of individual cells with specific functions arranged in a specific order. Skin, for example, comprises several thin layers of cells that are nourished by small blood vessels (Figure 2.2). Hair follicles

Epithelial cell   Lymphocyte

Nucleus
Cytoplasm

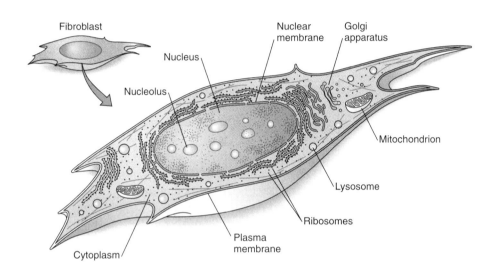

Fibroblast

Nucleus
Nucleolus

Nuclear
membrane

Golgi
apparatus

Mitochondrion

Lysosome

Ribosomes

Plasma
membrane

Cytoplasm

FIGURE 2.1
Prototype human cells.

EPIDERMIS

DERMIS

SUBCUTANEOUS
TISSUE

Hair

Pore

Germinal layer of
epidermis

Sebaceous gland

Nerve (sensory)

Sweat gland

Hair follicle

Blood vessel

Subcutaneous fat

Muscle fascia
Muscle

FIGURE 2.2
Layers of skin. Skin stem cells
constitute the germinal layer of
the epidermis.

15

and oil and sweat glands are other types of specialized cells that contribute to the function of skin. Collectively, the cells constitute the elastic body covering that is essential for life. Skin is an example of a tissue that contains a significant reservoir of stem cells and is a likely source of nuclei for nuclear transfer stem cells. Understanding stem cells, including those created by nuclear transplant technology, requires understanding fundamental properties of cells.

The Cell Cycle

When viewed through a microscope, a cell has two obvious stages: interphase, when the nucleus is intact, and mitosis, when the cell is dividing and the nuclei are not visible. For the most part, cell division brings about two cells, called daughter cells, which are identical to the original (Figure 2.3). Thus, it has long been appreciated that a cell is capable of completely duplicating itself. This fundamental observation in and of itself implies **inheritance** of all genetic traits from one cell to the next and lays the groundwork for genetic principles.

Inheritance: a gene passed on to offspring.

Gregor Mendel failed his examinations to become a monk in the 1850s but stayed at the monastery to conduct his famous white and purple pea blossom experiments (Figure 2.4), which established the basic genetic concepts of inheritance and dominance. Genetic traits are inherited equally from both parents, and some genes that define those traits are more likely to be expressed (dominant) than other genes (recessive). For example, the purple blossom gene for color is dominant over the white blossom gene for color. These same principles of genetics have been shown to be true for mammals, including humans, and indicate that all genetic information inherited from each parent must somehow be present in sperm and eggs. Gene expression is described in a later section in this chapter.

Once it was known that all cells contain all genes, arranged on chromosomes, it followed that mitosis must therefore yield two genetically identical daughter cells with two copies of each chromosome. Increased resolving power of microscopes confirmed chromosome pairing and their precise division during

FIGURE 2.3
Cell division. Cells increase in size, replicate their chromosomes, and divide into two new genetically identical cells.

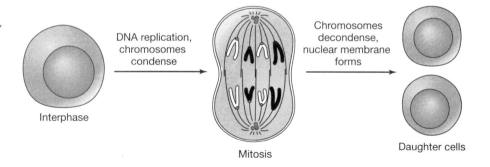

Interphase

DNA replication, chromosomes condense

Mitosis

Chromosomes decondense, nuclear membrane forms

Daughter cells

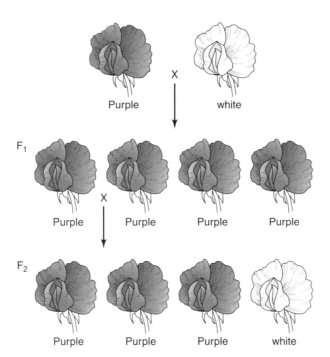

FIGURE 2.4
Gregor Mendel's pea blossom experiments. The color purple is dominant over white, so the first generation (F1) appears purple, but each carries a recessive gene for white. If two F1 plants are crossed, one-fourth of the next generation (F2) will have two purple genes, one-fourth will have two white genes, and one-half will have one purple and one white. Thus, three-fourths will appear purple.

mitosis (Figure 2.5). This suggested that the chromosomes contained within the nucleus might be genetic reservoirs. Interphase, the period of time during which a nucleus is visible but chromosomes are not organized, is therefore the time during which the chromosomes are duplicated.

Chromosomes are composed of the unique polymer, deoxyribonucleic acid (DNA), which encodes the genetic information. This discovery was made by Oswald Avery, Colin MacLeod, and Maclyn McCarty in 1944. Interestingly, the discovery was made in an attempt to understand the bacterium, *Pneumococcus*, that causes pneumonia. In the next decade, James Watson and Francis Crick published the structure of DNA as two highly organized polymers, ionically bonded together in a rope-like helical structure (Figure 2.6). Their predicted structure was based on data obtained from x-ray crystallography studies by Rosalind Franklin and Maurice Wilkins.

The fundamentally important discovery that inherited traits are encoded by DNA, which is a polymer of specific sequences of four molecules joined together to form an enormous double strand, led to an explosion of research on the mechanisms of chromosome duplication (Figure 2.6). The work continues today because all the details of replication of DNA are still not known. What is known is that duplication of all the DNA in each chromosome is necessary before the cells can divide. The period of DNA synthesis within interphase is termed **S phase** of the cell cycle.

"S" phase: the process of faithful duplication of each strand of DNA constituting each chromosome in a cell so as to yield two identical copies.

FIGURE 2.5
Human chromosome pairing
during mitosis.

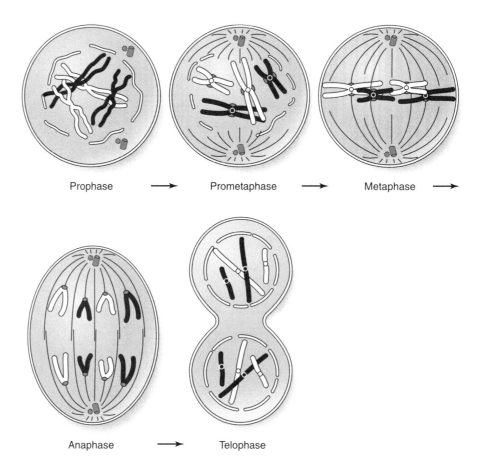

Prophase $\longrightarrow$ Prometaphase $\longrightarrow$ Metaphase $\longrightarrow$

Anaphase $\longrightarrow$ Telophase

The advent of radioactive forms of the individual DNA molecules allowed experiments to determine the duration of S phase and to provide clues about the process that leads to the faithful duplication of each DNA strand. It was discovered that there is a time interval after cell division and the appearance of the new nucleus but before DNA synthesis begins. This interval is defined as Gap 1 (G1). In addition, there is an interval of time between the end of S phase and the onset of chromosome formation. This interval is defined as Gap 2 (G2). It was also discovered that each strand of the DNA molecule serves as the template for a new strand. Thus, it is important to note that each new chromosome has one old strand of DNA and one new strand of DNA (Figure 2.6). This process is called semiconservative replication because one of the old strands is maintained. The newly replicated double strands represent a duplication of the chromosome.

The importance of faithful replication of DNA cannot be overstated. This is especially true for stem cells and cells in the early embryo because each cell is poised to have millions of progeny cells, which would all bear the replication defects. Controls on DNA replication are so complex that despite more than two

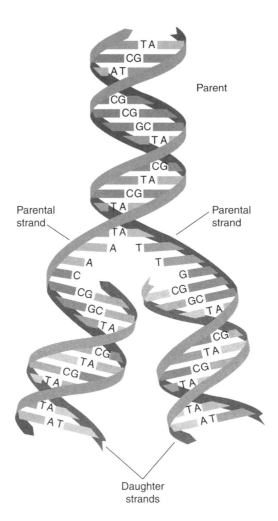

FIGURE 2.6
The double-stranded helical struc-ture of DNA and replication of both its strands. Newly replicated DNA strands contain one parent and one daughter strand.

Parent

Parental strand

Parental strand

Daughter strands

decades of research all the nuances are still not known, and no studies of DNA replication in human eggs have been reported. What is known for other cell types is that the process begins with the binding of a complex of six proteins to selected regions of the DNA that together form the **origin of replication.** This complex, in combination with a few other proteins, provides the framework for DNA polymerase to bind and begin the process of replication.

It may be important that for somatic cells, there are specified locations within the DNA for replication to begin, whereas for embryonic cells, there are many more origins of replication sites that account for the rapid cell cleavages. The huge size of the nucleus of early embryonic cells may provide the open scaf-folding necessary to support more origins of replication, thus allowing duplica-tion of genetic information as quickly as possible. Research on DNA replication in activated human eggs is urgently needed.

During G2, the cell prepares for division, termed *mitosis* or *M phase*. During early M phase, the DNA begins to rearrange itself according to individual chromosomes and discontinues its association with the scaffolding in the nucleus in preparation for mitosis. Each duplicated chromosome becomes compacted in a process termed *chromosome condensation* to describe the tightening of the DNA strands into coiled structures strong enough to withstand cell division without losing any genetic information. The duplicated chromosomes are termed *chromatids* because they stay joined at a central region, the centromere, which anchors the mitotic spindle, the strong protein fiber that pulls the chromatids apart at cell division.

Somatic: all cells and tissues of an organism except sperm and eggs and their precursor cells, termed *germ cells*.

Humans have 23 pairs of chromosomes—22 identical pairs termed **somatic chromosomes** and one pair of sex chromosomes, which is a pair of X chromosomes in the female but an X and Y chromosome in the male, designated for their appearance (Figure 2.7). One chromosome of each pair comes from the mother and the other from the father. The terms *somatic chromosomes* and *sex chromosomes* are descriptive, not functional. No particular chromosome gives rise directly to any particular organ, and the genes for testes do not reside strictly on the Y chromosome. However, genes with related functions are sometimes found clustered together in chromosomes. For example, genes that encode the proteins that define "self" displayed on the plasma membrane of cells cluster on the long arm of chromosome 6.

During mitosis, the chromatids are separated into the daughter cells, yielding new cells with two copies of each chromosome, one from the mother and one from the father (Figure 2.5).

A standard cell cycle is approximately 18 to 24 hours long, with a G1 of 7 to 9 hours, an S phase of 6 to 8 hours, a G2 of 2 to 4 hours, and an M phase of 1 to 2 hours (Figure 2.8). After cell division, some cells undergo additional changes in

FIGURE 2.8
Schematic of the phases of a
dividing cell.

function to carry out specific tasks. This change in cell function is termed *differentiation* and is usually accomplished in response to a specific factor that impacts the new daughter cell. For example, nerve growth factor is a protein that can bring about differentiation of stem cells into primitive neurons, as described in Chapter 1, Figure 1.2. Cells that undergo differentiation usually do not divide again. Those cells actually exit the cell cycle and enter a mitotically quiescent phase termed *Gap* 0 (G0). Most cells that enter G0 never divide again, but some, like stem cells, can be recruited from G0 to reenter G1. During this resting phase, the cell carries out its specific tasks as a differentiated cell in addition to housekeeping functions, such as respiration, and may grow a little, but it does not synthesize the proteins necessary to enter into a dividing cell cycle. The onset of cell division by most cells, including stem cells, requires a specific stimulus.

Stimulating Cell Division

The potato peeler accidentally removes a piece of finger skin. A cascade of emergency signals is immediately released by the damaged cells. A blood clot forms. Underlying skin cells (the dermis) knit together to seal the wound against fluid loss from below and infection from above. They turn on synthesis of keratins, the specialized proteins that toughen them to become surface cells (the epidermis). The stem cells beneath the dermis receive signals to begin to divide to replace the epidermis cells that are moving upward to become epidermis (Figure 2.2). The process takes at least 24 hours because the skin stem cells have been resting in G0 and must awaken to enter G1. The stem cells continue to produce daughter cells until the damaged area is repopulated.

A fundamentally important characteristic of stem cells is that only one, not both, of the daughter cells go on to differentiate; the other one must remain an undifferentiated stem cell to maintain the reservoir. Although there is still much to be learned, skin apparently has more stem cells than other body tissues, such as the heart, which is unable to replace damaged muscle cells within a few days. One possibility is that the heart does have stem cells, but their division is suppressed by other factors in the tissue that inhibit the cell cycle of all cells. Understanding all the factors involved in controlling cell division is an active area of research. Cell division and cell differentiation are guided and controlled by expression of specific genes, described briefly in the following section.

Gene Expression: Transcription and Translation

As discussed previously, genes are unique specific polymers of four molecules distinguished by side groups of adenosine (A), cytosine (C), guanosine (G), and thymidine (T). The sequence of the molecules in the polymer determines the nature of the protein specified by the gene. Genes are joined end to end in the long strands of DNA that constitute the chromosomes. Genes are highly variable in length and are distinguished from each other by specific "start" and "stop" sequences (Figure 2.9).

As depicted in Figure 2.10, the expression of genes into new proteins requires a series of well-coordinated steps. First, the specific sequence of A, C, G, and T in one strand of the DNA that composes the genes is transcribed, by a process termed **transcription**, into single-stranded RNA (ribonucleic acid) (Figure 2.11). Enzymes, termed RNA polymerases, synthesize the RNA to uniquely mimic the DNA sequences in the gene. The RNA strands contain molecules with three of the same four side groups (A, C, and G) and uridine (U) instead of thymidine. Three types of RNA are transcribed from individual genes: messenger RNA (mRNA), ribosomal RNA (rRNA), and transfer RNA (tRNA).

Transcription: the process of synthesizing messenger RNA (mRNA) from the gene sequence as the template.

FIGURE 2.9
Schematic of genes arranged end-to-end on chromosomes.

FIGURE 2.10
Flow scheme of transcription
and translation.

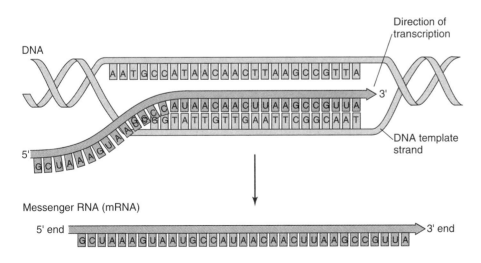

```
                    ┌──────────┐
                    │   DNA    │
                    └──────────┘
               ┌─────────────────────┐
               │   TRANSCRIPTION     │
               └─────────────────────┘
    ┌──────────┐   ┌──────────┐   ┌──────────┐
    │  rRNA    │   │  mRNA    │   │  tRNA    │
    │(ribosomal)│  │(messenger)│  │(transfer)│
    └──────────┘   └──────────┘   └──────────┘
                ┌──────────────┐
                │  Ribosome    │
                └──────────────┘
               ┌─────────────────────┐
               │    TRANSLATION      │
               └─────────────────────┘
                   ┌──────────┐
                   │ Protein  │
                   └──────────┘
```

mRNA is the template for protein synthesis. Proteins are polymers of molecules termed *amino acids*, of which there are approximately 20 types within
mammalian cells. The sequence in which the amino acids are joined together
determines the characteristics of the protein; the sequence of the four molecules
in the mRNA directs the sequence of amino acids in the protein encoded by the

FIGURE 2.11
Schematic of transcription.

Direction of transcription

DNA

AATGCCATAACAACTTAAGCCGTTA

3'

GCUAAAGUAAGGCCAUAACAACUUAAGCCGUUA
GGTATTGTTGAATTCGGCAAT

DNA template strand

5'

Messenger RNA (mRNA)

5' end → GCUAAAGUAAUGCCAUAACAACUUAAGCCGUUA → 3' end

gene, which is described in more detail below. rRNA is a structural molecule that combines with specialized proteins to form the ribosome, a globular structure fundamentally important to protein synthesis (Figure 2.12). tRNA is a small highly structured molecule that recognizes and transports the specific amino acids needed for the growing protein chain. Together, the RNAs constitute an efficient assembly line factory that produces proteins. The process is remarkably complicated and well controlled.

It begins with signals to synthesize RNA. A seemingly boundless repertoire of signals is involved in turning off and turning on RNA synthesis from genes. Hormones, growth factors, invasion by bacteria or viruses, starvation, gluttony, and sunshine are examples of stimuli for gene expression. Each stimulus excites an intracellular signal transduction pathway that, principally through a series of phosphorylations and dephosphorylations (the domino effect), in turn mobilize transcription factors to turn on or off gene expression in the nucleus. Some transcription factors are proteins that bind specifically to an area of the chromosome and expose the gene to the RNA polymerases. In addition to RNA polymerases, transcription involves a large complex of enzymes and other cofactors that bind to a specific region of the gene, termed the *promoter,* and systematically manufacture the RNAs, which then leave the nucleus for the cytoplasm. Multiple RNAs can be transcribed from the same gene at the same time. rRNAs combine with ribosomal proteins in the cytoplasm. The tRNAs coil up on themselves and remain free spirits to search the cytoplasm for their specific amino acid.

mRNA slides between the two subunits of the ribosomes (Figure 2.12), exposing its sequence in groups of three molecules, the **triplet codon,** which defines an individual animo acid. For example, the sequence GGU specifies the amino acid glycine, the sequence AUU specifies isoleucine, the sequence GUU specifies valine, and GAA specifies glutamic acid. Together, the sequence GGUAUUGUUGAA specifies the amino acid chain glycine-isoleucine-valine-glutamic acid, the first four amino acids in the hormone insulin. The process of reading the mRNA sequence and synthesizing the corresponding protein is termed *translation.*

The triplet attracts its amino acid by binding with a specific tRNA that, in turn, has found its amino acid. The amino acid is brought into the ribosomal–mRNA complex and is positioned as the next amino acid in the chain (Figure 2.12). It is chemically bonded to the growing chain like the next piece in a jigsaw puzzle. It all happens very fast. Proteins are synthesized at the rate of 20 amino acids per second. Thus, one molecule of insulin can be synthesized in approximately 3 seconds. Each mRNA can simultaneously direct the synthesis of multiple growing protein strands, and each gene can transcribe multiple mRNAs simultaneously. Thus, millions of copies of proteins can be produced from one gene sequence in a matter of minutes, and production can be halted just as fast. The speed, precision, and responsiveness of gene expression are awe-inspiring cellular assets.

In no cell is it more finely tuned than in the egg. As described in Chapters 3 and 4, unlike dividing somatic cells that synthesize and degrade proteins, as

Triplet codon: the sequence of three bases in mRNA that define an amino acid in the encoded protein; the mRNA triplet codon aligns with the specific transfer RNA (tRNA) bound to the corresponding amino acid.

FIGURE 2.12
Schematic of translation.

Gene Expression: Transcription and Translation

directed by either cell cycle machinery or specific cellular functions, the egg synthesizes large stockpiles that must be deployed as needed. Ribosomes and ribosomal precursors are stockpiled in massive quantities by the egg. Large stores of mRNAs, termed *maternal messages* in eggs, such as cMos, which is essential for metaphase II arrest (see sidebar, Chapter 3), are also stockpiled. Other cellular components essential for rapid and accurate early cleavages are stored as proteins, such as the DNA polymerase detected in relatively high concentration in metaphase II–arrested mouse eggs. The systematic deployment of these stockpiles for the multiple cycles of DNA replication and cleavage is a property of eggs that remains poorly understood, especially with respect to human eggs.

Molecular Biology of the Cell Cycle

Early work, which suggested that factors outside the nucleus, in the cell cytoplasm, controlled the cell cycle, came from studies of frog eggs, as described in Chapter 3. During the same period of time, three scientists—Lee Hartwell, Paul Nurse, and Tim Hunt—working with yeast and sea urchin eggs discovered fundamentally important cytoplasmic controls on cell division. The three scientists shared the Nobel Prize in 2001 for their discoveries (see sidebar).

Progression through the cell cycle is now known to be regulated by a specific set of genes that encode proteins that can add phosphate groups to other proteins (Figure 2.13). Such proteins are enzymes called **kinases**. Enzymes are proteins that bring about chemical reactions at body temperature that would otherwise require heat. Phosphate groups have a strong ionic charge that either attracts or repels other groups and, in this way, changes the shape of proteins. Sometimes changing their shape causes enzymes to become more active, sometimes less active.

Two classes of proteins encoded by genes important to the cell cycle are cyclins and cyclin-dependent kinases (Cdks). All cells with nuclei, from yeast to

Kinase: a protein (enzyme) capable of bringing about the addition of a phosphate group ($PO_4^=$) to an amino acid.

FIGURE 2.13
Protein phosphorylation and dephosphorylation. One of the phosphate groups from adenosine triphosphate is added to the hydroxyl group of serine, threonine, or tyrosine by a kinase enzyme; the phosphate group is subject to removal by a different enzyme, a phosphatase.

Cell Cycle Regulators
Lee Hartwell, Paul Nurse, and Tim Hunt

In 2001, three scientists—Lee Hartwell, Paul Nurse, and Tim Hunt—shared the Nobel Prize in Physiology or Medicine "for their discoveries of key regulators of the cell cycle." In his Nobel Prize Presentation Speech, Professor Anders Zetterberg noted that

> All living organisms on earth are descended from an ancestral cell that appeared about three billion years ago, and which has undergone an unbroken series of cell divisions since then. Each human being began life as one single cell—a cell that divided repeatedly to give rise to all one hundred thousand billion cells that we consist of. Every second, millions of cells divide in our body.

Dr. Hartwell did his undergraduate work at California Institute of Technology and his Ph.D. studies at the Massachusetts Institute of Technology. As a young researcher at the University of California, he began to use baker's yeast as his model organism. Although a microorganism, yeast has a nucleus that contains the yeast's chromosomes. The advantage of yeast as an experimental model is that it has a short cell cycle time. The short cell cycle time speeds up the formation of mutant organisms. By carefully characterizing mutant organisms, some of which were unable to divide at slightly higher than normal temperatures, Lee Hartwell described dozens of genes important to cell division. He termed these genes *cell division cycle* (CDC) genes. He noticed that the cell cycle only proceeded forward if the genes were expressed in a specific order. From this observation, he formulated the concept of cell cycle checkpoints. One gene, *CDC28*, proved to be essential for the initiation of each cell cycle.

Many checkpoints are in G1. For example, cells must grow to an appropriate size before initiating DNA synthesis. This requires appropriate nutrients and growth factors. Dr. Paul Nurse also used yeast for his cell cycle studies, but not baker's yeast. Dr. Nurse completed his undergraduate studies at the University of Birmingham and his graduate studies for his Ph.D. at the University of East Anglia in England. Dr. Nurse discovered that specific mutations in cell division cycle gene number two, Cdc2, led to cells that either did not divide or divided too early. He hypothesized that Cdc2 was a principal regulator in the cell cycle, and, in fact, controlled the cell cycle. By incorporating human genes into mutant yeast, he proved that the human gene analogous to the yeast gene (a homolog) could perform the correct function in yeast. Thus, Cdc2 in the yeast used by Dr. Nurse was the same gene as *CDC28* in baker's yeast, and the same gene as human Cdc2. Remarkably, a gene responsible for initiating the cell cycle in baker's yeast had been conserved for more than one billion years of evolution, from one-celled organisms to humans. This illustrates a fundamental biologic principle: If it works, don't mess with it.

Dr. Tim Hunt completed both his undergraduate and graduate studies at the University of Cambridge in England. Dr. Hunt worked with sea urchin eggs as a model organism. In 1982, he discovered another gene that was expressed immediately before cell division but not expressed at all immediately after cell division. He named the gene cyclin and discovered its existence in other types of cells as well. Soon thereafter numerous cyclin genes were discovered, and the proteins they encode form complexes with the proteins discovered by Dr. Nurse and Dr. Hartwell. These fundamentally important discoveries paved the way for major advances in understanding cellular controls.

Cell Cycle Phase	Cyclin	Cdk
G1	A, D1, D2, D3, E	Cdk2 (Cdc2)
	D1, D2, D3	Cdk4
	D2, D2, D3	Cdk5
S	A, B	Cdk1
G2 to M	A, B	Cdk1

TABLE 2.1 Mammalian Cell Cycle Regulators

human, contain genes that encode very similar protein families (Table 2.1). As the names imply, Cdks are not active until they bind to their respective cyclin, a protein that is not an enzyme on its own. Cyclin is so named because it is synthesized at specific stages of the cell cycle. It binds to its specific Cdk and brings about a change in the Cdk structure, leading to activation of kinase activity.

There are several related, but different, cyclins. For example, cyclin E activates Cdk2, which is responsible for the transition of the cell from G1 to S phase. Because cyclin E is degraded immediately after it activates Cdk2, it must be synthesized by the cell during each G1 phase. The cellular controls that turn on and turn off synthesis of the proteins encoded by the cyclin genes are clearly fundamentally important to whether or not a cell is capable of cell division. This is an active area of investigation.

Similarly, cyclins A and B regulate the activity of another kinase, Cdk1, which is essential for the transition from S phase to G2. The observation that separate sets of cyclins and Cdks function at defined points in the cell cycle established the possibility of checkpoints that could stop the progression through the cell cycle if an error has occurred. For example, an important checkpoint occurs in G1 before the initiation of S phase. DNA damage is repaired before replication of the DNA begins. A protein, p53, mediates the G1 checkpoint. Interestingly, many human cancers contain mutations in the *p53* gene that lead to synthesis of a defective p53 protein that is not capable of mediating the checkpoint. This, therefore, allows cells with DNA damage to initiate S phase. In this way, defective cells characteristic of cancerous tumors are allowed to accumulate. Such cells have escaped the natural checkpoints for cell division and also usually fail to carry out the normal functions of the tissue in which they arose.

Another important checkpoint is at the end of S phase. DNA replication requires an enormous complex of proteins, including DNA polymerases and other accessory proteins. They are assembled during G1 and disassembled during late S phase. In this way, the cell maintains strict control of the process of DNA synthesis. As seen in later chapters, an exception to this rigid control exists in eggs

FIGURE 2.14
Origin of replication complex. DNA replication begins at specific locations chosen by proteins that interact with each other and with DNA strands.

Helicase

DNA polymerase complex

DNA ligase

Gyrase (topoisomerase II)

RNA primase complex (primosome)

Single-strand binding proteins (SSB proteins)

DNA polymerase complex

and early embryos, which exhibit unusual cell cycles, the reasons for which remain a mystery. It is obviously critically important to the cell line that mitosis does not begin until DNA replication is completed. Moreover, should something happen to the cell during S phase, such as exposure to x-rays or ultraviolet light, damage to the DNA would need to be repaired before S phase could be completed. In circumstances where the damage cannot be repaired, cell death frequently occurs as a result of stimulation of specific cell death pathway enzymes.

DNA replication begins at specific locations in the chromosomes, approximately 200 or 300 thousand bases apart in mammalian cells. The site is chosen by specific proteins, usually six in number, that compose the origin of replication complex (ORC) (Figure 2.14). The proteins form a complex, with each member having a specific function. For example, ORC1, ORC4, and ORC5 proteins bind ATP. It is possible that the ORC remains associated with chromatin at all phases of the cell cycle but promotes binding of the remaining proteins required for DNA synthesis only at the G1/S transition. This is an active area of research in mammalian cells.

Once DNA replication is complete, the ORC undergoes alterations that inhibit further DNA synthesis, and the cell progresses to G2. At this stage, the cell contains two copies of each chromosome, a state referred to as 4N. During this phase of the cell cycle, the cell continues to increase in size and prepares for cell division. Progression from G2 to M phase requires activation of Cdk2 (sometimes denoted Cdc2) by binding to cyclin B, which leads to an activity originally identified as maturation promoting factor (MPF) in frog eggs, described in Chapter 3. Cyclin B synthesis begins during S phase. As it is synthesized, it forms a complex with Cdk2. The complex is immediately phosphorylated at three amino acids on Cdc2 (threonine 14, tyrosine 15, and threonine 161) to maintain its inactive state (Figure 2.15). As discussed previously, adding phosphate groups (phosphorylation) can either activate or inactivate enzymes. Enzymes that remove phosphate groups are termed **phosphatases** (Figure 2.13). Cdc25 is the phosphatase that brings about the activation of the cyclin B–Cdc2 complex

Phosphatase: a protein (enzyme) capable of removing a phosphate group ($PO_4^=$) from an amino acid.

FIGURE 2.15
Regulation of maturation
promoting factor.

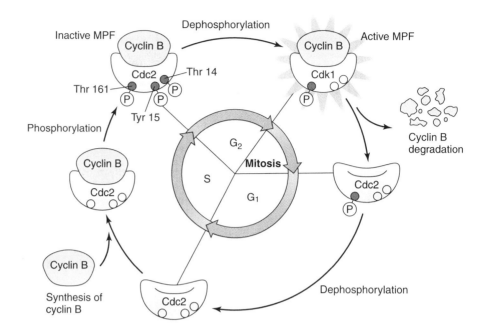

(MPF) by removing the phosphate groups from threonine 14 and tyrosine 15. The active complex is a kinase that phosphorylates several proteins essential to M phase, including other kinases and structural proteins, such as histone 1, the protein that complexes with DNA and whose rearrangement is necessary for the formation of chromosomes, the nuclear lamins, which lead to dissolution of the nuclear membrane, and microtubules, which lead to the formation of the mitotic spindle.

A checkpoint also exists during mitosis to ensure equal division of the chromosomes into each daughter cell. Once the chromosomes are lined up on the mitotic spindle, MPF must be inactivated to allow mitosis to proceed. This is accomplished by degradation of cyclin B by a pathway specifically activated by MPF at the initiation of M phase. Thus, MPF ensures its own finite lifetime specifically to orchestrate metaphase. The decline in MPF activity allows the formation of two new nuclei and a division of the cell cytoplasm and plasma membrane.

Summary

Tissues and organs are composed of individual cells with specific functions. Fully specialized cells, termed *differentiated,* do not divide. Replacement of defective or dying cells depends on a reservoir of stem cells, present in some, but not all, tissues.

Stem cells divide into daughter cells when called on. Duplication of a cell into two daughter cells is under the control of highly conserved genes. Their function is to optimize the possibility that two daughter cells, identical to the parent cell, will be formed. The process involves growth of the cell in size, duplication of the cellular organelles, synthesis of the large complex of DNA polymerases and accessory proteins responsible for faithful replication of the DNA, DNA synthesis, the formation of chromosomes and their accurate division into the two halves, and cell division, followed by the formation of new nuclei. In no cell are these processes more critical than in the stem cell, which serves as a life-long reservoir of cells capable of self-renewal.

All cell functions are controlled by specific genes. Genetic information is encoded by a series of four molecules contained in unique sequences in the long molecules of DNA that compose the chromosomes within the nucleus. A copy of the genetic information, termed *messenger RNA* (mRNA), exits the nucleus and serves as the template for the precise organization of the amino acids that make up the protein encoded by the gene.

Cell division is divided into four parts that reflect the presence or absence of the nucleus that contains the genetic information. Cell division is termed *mitosis*. Once two new daughter cells are formed, the period of time before the initiation of DNA synthesis is termed *Gap 1* (G1). The period of DNA synthesis is termed synthesis (S) phase. Once the DNA is completely replicated, the period of time until mitosis is termed *Gap 2* (G2). And so the cycle repeats itself.

Review Questions (True or False)

1. All cells arise from other cells.
2. Human cells have 46 chromosomes, half of which are inherited from the father and half from the mother.
3. Chromosomes are composed of DNA which is a double-stranded polymer that is duplicated during S phase.
4. New DNA synthesis must result in a faithful copy of each chromosome for each new daughter cell. This is accomplished by using each strand of the double-stranded polymer as the template on which to synthesize the new strand.
5. The intermediate molecule between genes and their protein products is messenger RNA.
6. The addition of a phosphate group is the only way to activate proteins to carry out their normal cell functions.
7. DNA synthesis begins at one place on each chromosome.

ADDITIONAL READINGS

DePamphilis, M. L. (2005). Cell cycle dependent regulation of the origin recognition complex. *Cell Cycle* 4: 70–79.

Elledge, S. J. (1996). Cell cycle checkpoints: preventing an identity crisis. *Science* 274: 1664–1672.

Oostendorp, R., Audet, T., and Eaves, C. (2000). High-resolution tracking of cell division suggests similar cell cycle kinetics of hematopoietic stem cells stimulated in vitro and in vivo. *Blood* 95: 855–862.

Pal, S. K., Torry, D., Serta, R., Crowell, R. C., Seibel, M. M., Cooper, G. M., and Kiessling, A. A. (1994). Expression and potential function of the c-mos protooncogene in human eggs. *Fertil Steril* 61: 496–503.

Weinert, T. (1998). DNA damage and checkpoint pathways: molecular anatomy and interactions with repair. *Cell* 94: 555–558.

PART

II

Egg-Specific Functions

CHAPTER 3

The Egg

When you start with a portrait and search for a pure form, a clear volume, through successive eliminations, you arrive inevitably at the egg.

Pablo Picasso

OVERVIEW

The egg is wondrous. One of its most noteworthy features is size. The fully grown human egg has a diameter of 110 to 120 microns, yielding a volume of approximately 900,000 cubic microns. In comparison, a human white blood cell has a diameter of about 20 microns, yielding a volume of approximately 4,000 cubic microns. The volume of a sperm head, in contrast, is approximately 250 cubic microns. Thus, an egg is nearly 250 times larger than a white blood cell and nearly 4,000 times larger than a sperm head.

The nucleus of the egg is also huge. One of the first structures noted by early microscopists, it was called the **germinal vesicle** (Figure 3.1), a term still in common use today. It is uniquely characterized by a large prominent nucleolus, which is the organizing region for ribosomal RNAs. Together, the germinal vesicle with its large nucleolus resembles a giant belly button in the center of the egg. With a diameter on the order of 50 microns, the volume of the germinal vesicle is approximately 65,000 cubic microns, more than 10 times the volume of a white blood cell. This huge nucleus provides ample open scaffolding for egg chromosomes that may play an important role in the ease of expression of egg (oocyte) genes.

The egg also has an unusual and unique cell cycle. Given the importance of reproduction to the species, it is reasonable to assume there are multiple safeguards to ensure an adequate supply of eggs with the potential to give rise to new offspring. Many aspects of egg development remain mysterious, however, principally because unlike other cells, laboratory methods have not been developed to support the life cycle of eggs in culture. This fundamentally important

Germinal vesicle: the large nucleus of an oocyte.

FIGURE 3.1
Germinal vesicle stage human egg.

roadblock is the largest hurdle faced by nuclear transplant stem cell scientists. A report of the production of mouse eggs from mouse embryonic stem cells in culture generated much excitement that mammalian egg biology could be studied in greater detail in the laboratory, but the findings have been difficult to repeat and have not been reported by other laboratories. In the absence of a laboratory culture system to generate oocytes, the sole source of human eggs at the time of this writing is surgical recovery from the ovaries of normal, healthy, adult women.

Egg Growth

The supply of eggs in the ovaries of each woman is established early in her fetal development. The development of the ovary is less well studied in humans than in mice; much of the information about human ovarian development is therefore extrapolated from mouse and from some monkey models. Although the timing is not known with certainty for humans, as described in Chapter 9, at some point before organs are fully formed, the precursors to what will become either eggs

Germ cells: general term for sperm and eggs and their precursor cells.

or sperm (collectively termed **germ cells**) gather together in an area termed the gonadal ridge. The germ cell precursors are, in fact, stem cells for sperm and eggs. A difference between stem cells for sperm and stem cells for eggs is that egg stem cells contain no Y chromosome. As described in Chapter 2, humans have 23 pairs of chromosomes in each cell—22 somatic chromosomes and 1 pair of sex chromosomes. In the female, the sex chromosomes are a pair of X chromosomes (for their symmetric shape), and in the male the sex chromosome pair has one X and one Y chromosome (see Figure 2.7).

The presence of a Y chromosome leads to sufficient production of testosterone by the developing male fetus to cause the gonadal ridge to develop into a testis and support the development of other male reproductive tract organs, such as the epididymis and seminal vesicles. In the absence of a Y chromosome, an ovary forms from the gonadal ridge along with other female reproductive tract tissues, such as oviduct (fallopian tubes) and uterus, described in Chapter 9. Egg stem cells, oogonia, divide continuously in the fetal ovary during early development. Oogonia are not significantly larger than somatic cells. For reasons not fully understood, oogonia stop dividing in the fetal ovary during the second trimester of fetal development and become primary oocytes. Primary oocytes are arrested in late G2 of the cell cycle. This is a highly unusual cell cycle arrest point, and the reproductive strategy behind this arrest is not entirely certain. In this state, they have twice the normal amount of DNA (tetraploid) and are said to be in the prophase of meiosis. This is the stage at which chromosome crossover can occur, a process most thoroughly studied in yeast.

Each cell in the developing fetus has one chromosome from the father and one chromosome from the mother. This is the genetic paradigm first predicted by Mendel's pea blossom experiments, described in Chapter 2. Chromosome crossover allows balanced exchange of DNA sequences (and therefore genes) between two like chromosomes. Although they are not, strictly speaking, chromosomes during prophase, the DNA strands that compose each chromosome somehow recognize each other and matching sections of the DNA line up in a close orientation. They are positioned such that the DNA strands are spliced open and reunited with a strand from the paired chromosome (Figure 3.2). Because there are four copies of each chromosome, there are multiple opportunities for chromosome crossover in the primary oocyte. This process leads to new chromosomes that contain genes from both the father and the mother. Because the process of chromosomal crossover is thought to be relatively random, each egg that completes prophase of meiosis could theoretically have a unique set of chromosomes.

The primary oocytes do not resume growth and maturation for at least a decade, at the time the girl enters puberty and a regular monthly pattern of egg maturation and release begins. It is not known with certainty if chromosome crossover is continuous in human oocytes throughout the long prophase of meiosis. It is known, however, that as soon as a population of oocytes is established, that population begins a decline in numbers that eventually culminates in menopause. Infant girls are born with on the order of 1 million primary oocytes. By the time she is approximately 50 years of age, the oocytes are all gone. The best estimates are that

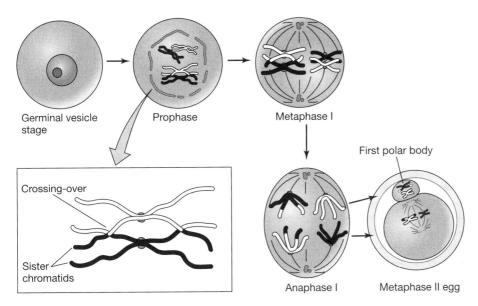

FIGURE 3.2
Chromosome crossing over.

Germinal vesicle
stage

Prophase

Metaphase I

First polar body

Crossing-over

Sister
chromatids

Anaphase I

Metaphase II egg

after birth, oocyte death is essentially linear. This means that on the order of 20,000 eggs die each year, including the dozen or so which are ovulated.

Whether or not the eggs recruited for ovulation are selected from the remaining healthy cohort, rather than from the cohort that has entered the death pathway, is not known with certainty. As discussed later in this chapter, many more human eggs can be fertilized and initiate development than actually give rise to offspring. This suggests that defective oocytes may be included in the process of ovulation. There is ample evidence that the failure of fertilized eggs to develop into offspring is not due to failure to develop into multiple types of cells but failure to develop functioning organs, such as the heart. This suggests that many more eggs may have the potential to give rise to stem cells after nuclear transplant (ovasomes, see Chapters 7 and 10) than have the potential to give rise to infants. The human and medical value of being able to distinguish between cleaving eggs with the full potential to give rise to offspring and those with limited potential cannot be overstated. Research in this area has been essentially eliminated by the ruling of the U.S. Congress that ". . . such research is valuable, but will not be funded by taxpayer dollars."

Oocyte Meiosis

As described previously, oogonia stop dividing in the fetal ovary. This is in contrast to spermatogonia, which continue to divide throughout the life of the adult male. The oogonia undergo one final round of DNA synthesis, S phase, enter G2, grow larger, and enter the prophase of meiosis. As primary oocytes they are approximately 35 microns in diameter, completely surrounded by a layer of

FIGURE 3.3
Hormonal communication between
the brain and the ovary.

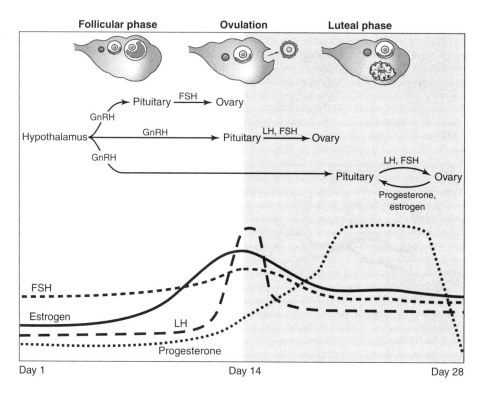

Ovulation: the process of
release of an egg from the
ovary.

granulosa cells (termed *nurse cells*), specialized hormone-producing cells in the ovary that play a major supportive role in oocyte growth, maturation, and eventual release from the ovary at ovulation.

Maturation of a primary oocyte in preparation for **ovulation** is intimately tied to the monthly menstrual cycle. It is the result of wonderfully orchestrated communication between the ovary and two glands in the brain, the hypothalamus and the pituitary (Figure 3.3). The hypothalamus and pituitary collaborate on a number of important body functions in both sexes. The hypothalamus produces protein hormones whose target is the pituitary. The pituitary gland in turn synthesizes and releases more protein hormones. For example, the hypothalamus synthesizes and releases growth hormone-releasing hormone, which stimulates the pituitary gland to produce growth hormone to stimulate growth and development in children, including the long bones of the legs and arms. It seems as if once growth is essentially accomplished, the hypothalamus and pituitary turn their attention to the ovaries and the testes, collectively called gonads. The hypothalamus synthesizes and releases gonadotropin-releasing hormone (GnRH), which stimulates the pituitary to release hormones to the bloodstream that stimulate the testes in boys and the ovaries in girls. The pituitary hormones are thus termed *gonadotropins*. They are released from the pituitary into the bloodstream in well-defined pulses throughout the day and night. The pulsatile hormone release persists until menopause in women and forever in men.

One of the gonadotropins is follicle-stimulating hormone (FSH), which stimulates sperm production in the testis and growth and meiosis in primary oocytes in the ovary (Figure 3.3). It is not certain how eggs are selected for ovulation each month. It is theorized that several may be selected but that one responds more vigorously than the others and eventually becomes dominant. FSH stimulates the granulosa cells surrounding the primary oocyte to divide and secrete estrogen. Estrogen in turn stimulates increased expression of receptors for FSH on the surface of the granulosa cells, thus capturing more FSH each time it is released into the bloodstream by the pituitary. In response, the oocyte resumes growth and expression of oocyte-specific proteins, including zona pellucida proteins, which form a protective coat around the giant cell (Figure 3.1). The zona pellucida plays several roles, as described in Chapter 6.

The cyclic recruitment of a primary oocyte to undergo maturation for ovulation that month begins within 1 to 2 days of the onset of menstruation, which signals that no pregnancy has occurred. In the absence of pregnancy, the granulosa cells in the ovary essentially cease synthesis of the steroid hormones, estrogen and progesterone. The decrease in steroid hormones stimulates the hypothalamus and pituitary to resume their hormone synthesis. The cycle thus begins anew in response to FSH release by the pituitary. Approximately 10 days after the onset of menstruation, FSH release has stimulated sufficient estrogen production by the granulosa cells surrounding the primary oocyte to in turn stimulate the development of a fluid-filled estrogen-rich sac termed an *ovarian follicle*. The advent of the ability to computer render echoes from low-energy ultrasound waves has given rise to equipment that can display a computer-generated picture of the developing follicle without apparently damaging the egg. A probe that generates low-energy ultrasound waves from its tip and senses their echoes is placed either on the abdomen or in the vagina, directed toward the ovary, and the growing follicle is visualized (Figure 3.4).

Within the hormone- and nutrient-rich environment of the follicle, the oocyte cytoplasm stockpiles components it will need to resume development after being fertilized by sperm. The stockpiling effort is assisted by projections of the granulosa cells, which extend through the zona pellucida into the egg cytoplasm. Numerous studies have shown that the granulosa cells thus serve as conduits for small molecules, such as amino acids and nucleic acids, which are shuttled directly from the follicular fluid into the egg cytoplasm. Once oocyte growth resumes, larger molecules, such as the enzyme horseradish peroxidase, are also shuttled into the egg's cytoplasm. The oocyte also synthesizes and stockpiles proteins and messenger RNAs (mRNAs) as it completes its growth. An unusual feature of eggs that is not understood is the inactive state of oocyte mitochondria. Although they are abundant within the egg cytoplasm, they appear not to be actively involved in metabolism, a feature that distinguishes them from those in dividing cells. The role of maternal mitochondria in oocyte metabolism and potential for development is an active area of research.

FIGURE 3.4
Growing ovarian follicles visualized
by ultrasound. The dashed lines
indicate measurement in
centimeters.

Estrogen produced by the granulosa cells not only acts locally to stimulate the egg and other cells in the growing follicle, it also enters the bloodstream and stimulates the pituitary to decrease production of FSH and stimulate release of another gonadotropin, luteinizing hormone (LH). By approximately day 13 of the menstrual cycle, LH pulses reach the same height as FSH pulses, which begin to decline (Figure 3.3). This circumstance initiates a dramatic cascade of responses in the follicle that ultimately leads to release of an egg poised for activation by sperm.

LH transforms the granulosa cells from producing estrogen to producing progesterone, and they stop dividing. Perhaps in response to the rise in progesterone, which is known to stimulate meiosis in frog eggs, the oocyte rapidly resumes meiosis. Protein and nucleic acid synthesis ceases. The huge germinal vesicle migrates toward the edge of the oocyte. Chromosome condensation begins, and the germinal vesicle membrane dissolves in preparation for the formation of the first metaphase plate, the lineup of chromosomes attached to spindle fibers that, in the egg, appears adjacent to the plasma membrane rather than in the center of the cell as in somatic cells. Almost as soon as the metaphase plate is formed, a unique unequal cell division occurs, which results in the production of a small cell, the polar body, approximately the size of a somatic cell, which contains a complete set of chromosomes (Figure 3.5). In contrast to mitosis (see Figure 2.5), whole chromosomes, rather than individual chromatids, are segregated. This is termed *meiosis I*. Because of the crossing over between chromosomes that occurred in the prophase of meiosis, the egg and the polar body have both maternal and paternal genes rearranged on the chromosomes.

After the unequal cell division, in contrast to all other cell cycles, the nuclear membrane does not reform around the remaining chromosomes. They immediately

FIGURE 3.5
Egg arrested at metaphase II,
surrounded by cumulus cells. PB,
polar body; MP, metaphase plate.

undergo a rearrangement, which results in a second metaphase plate that contains the normal number of chromosomes of a somatic cell. Thus, separation of the chromatids at meiosis II yields two sets of single chromosomes (termed *chromatids*) rather than chromosome pairs.

Metaphase plate: the newly duplicated chromosomes separated opposite from each other by protein fibers.

During this time, the granulosa cells immediately surrounding the egg retract their projections back through the zona pellucida and synthesize and secrete hyaluronic acid, a viscous protein through which sperm must propel themselves. Their new appearance around the egg is less dense and termed the *corona radiata*. Because these granulosa cells have remained attached to the egg itself, instead of lining the walls of the follicle, they are termed cumulus cells (Figure 3.5). The follicular fluid has changed from estrogen rich to progesterone rich, and progesterone levels rise within peripheral blood. Ovulation is imminent. Within 1 to 2 days of the peak of LH, the wall of the follicle ruptures, releasing the egg–cumulus complex, approximately 1 centimeter in diameter. The complex is swept into the fallopian tube by pulsating cilia, developed within 2 or 3 days of ovulation to direct the egg toward the uterus.

Despite this elaborate highly orchestrated preparation, the huge cell arrested at metaphase II is very fragile. If not activated within 1 to 2 days, the egg will pass unnoticed into the uterus and perish. As described previously, even if it is fertilized by a sperm, approximately three of four eggs do not have the capacity to give rise to offspring. Those who believe that life begins at fertilization must accept the fact that 75% of such newly formed "lives" naturally perish. Understanding what leads to developmental success or failure is clearly an urgently needed area of research.

Assisted Reproductive Technologies

By the mid-1980s, a variety of medical treatments for infertility had been developed, discussed further in Chapter 4. One major advance was the development of microsurgical techniques to bypass and repair damaged fallopian tubes that

IVF: in vitro fertilization, fertilization of an egg by a sperm in a laboratory tissue culture dish.

blocked the passage of the egg from the ovary into the uterus. Another major advance was the discovery that the urine of menopausal women contains relatively high concentrations of gonadotropins. When all the eggs in the ovary have died, the cyclic production of estrogen and progesterone ceases and there is no feedback to the hypothalamus and pituitary to block the release of FSH and LH. As a consequence, menopausal women have high circulating levels of FSH and LH for a few years, some of which pass through the kidney with urine. An Italian company developed the laboratory methods for purifying gonadotropins from urine collected in convents by nuns. The availability of pharmacologic doses of such gonadotropins provided the opportunity to treat women whose infertility resulted from their own hormone imbalances. It also provided the opportunity to increase the levels of gonadotropins in hormonally normal women to stimulate more than one egg to resume maturation each month, thus paving the way to increase the number of eggs available for fertilization in laboratories. Such **in vitro fertilization (IVF)** procedures have become the standard of care for a variety of infertility conditions, including low sperm counts.

As listed in Table 1.2, approximately 100,000 women undergo egg collection each year in the United States. The goal is to stimulate final maturation of 10 to 20 oocytes in concert. This allows several eggs to be fertilized and to begin development, as described in Chapters 5 and 6. The most normal-appearing embryos are then selected for transfer to the woman's uterus for possible gestation.

To help ensure that the eggs are all at the same stage of maturation at the time of collection, hormones are administered to block the pituitary stimulation of the ovary. There are at least two possible approaches. One is birth control pills. They are combinations of estrogen and progesterone that provide artificial feedback to the pituitary to stop releasing FSH. After a few weeks the ovary is quiet; no eggs have resumed maturation. A second approach is a chemical that actually stimulates an outpouring of GnRH from the hypothalamus, called a GnRH agonist. After 2 or 3 days, the over-stimulated hypothalamus actually shuts down its release of GnRH, a circumstance termed *down-regulation*. This leads to no FSH production by the pituitary. After several days, the ovary is quiet.

Once the ovary is quiet, recruitment of several oocytes may be accomplished by administration of relatively high doses of FSH. This clinical treatment is termed *controlled ovarian hyperstimulation* (COH). The urinary hormone originally developed by the Italian company has now been replaced by recombinant hormone synthesized in pharmaceutical laboratories. The artificial stimulation of the ovary brings about maturation of several oocytes in concert and the simultaneous division of the granulosa cells to several million per egg. They produce estrogen, which may play an important role in the final stages of egg maturation as well as prepare the lining of the uterus to receive the fertilized egg.

For the fertility clinic to plan the oocyte collection before spontaneous ovulation occurs, the mid-cycle surge of LH by the pituitary must also be blocked. This could be accomplished by continual administration of birth control pills, but that would also inhibit the ovarian response to the artificial FSH, so the birth control pills must be stopped before the COH is begun. The agonist does not appear to inhibit COH, so it may be continued throughout the period of COH.

Alternatively, a chemical was developed that inhibited, rather than stimulated, hormone production by the hypothalamus. It is termed an *antagonist* and is administered after the COH is begun specifically to inhibit spontaneous LH release by the pituitary. Importantly, despite the pharmacologic intervention, the eggs recruited do not necessarily constitute a synchronous cohort, and the growing follicles have a broad range of sizes (Figure 3.4). The COH is continued for 10 to 12 days, approximately the same number of days of stimulation as occurs during a natural menstrual cycle. When the developing follicles have grown to approximately 2 centimeters in diameter and appear to be producing sufficient estrogen, an artificial surge of LH is administered by a single injection of a high concentration of gonadotropin. Nearly all eggs that have responded to the COH resume meiosis in response to a single injection of a gonadotropin that mimics the natural LH surge (Figure 3.3). In this regard, it is important to note that the capacity to resume meiosis II may precede full maturation of oocyte cytoplasm.

Approximately 34 hours after administration of the artificial LH surge, eggs can be collected directly from the ovarian follicles with the aid of ultrasound visualization. This timing is close to the LH-induced rupture of the follicle to allow meiosis to proceed within the follicle as in normal ovulatory cycles. Numerous clinical studies have revealed that more than 80% of eggs collected from women undergoing COH have reached metaphase II, undergo successful fertilization by sperm, and initiate cleavage divisions (described in Chapter 6), but fewer than 10% of the fertilized eggs have the capacity to develop to a fetus. Unfortunately, at this time there are no definitive indicators to identify the cleaving eggs that have the potential to develop into a fetus. As a consequence, most U.S. infertility clinics allow two or more early cleaving eggs to be returned to the uterus of the infertile woman for gestation on the hope that at least one will continue to develop. Although this practice has successfully improved the pregnancy rate per egg collection, it has also resulted in an epidemic of twins and triplets that are high-risk pregnancies. To reduce the incidence of multiple births, some infertility programs allow the fertilized eggs to develop in the laboratory for 5 or 6 days, at which time the most robust embryos develop to the blastocyst stage (see Chapter 8), one or two of which are selected for transfer to the uterus. This strategy depends on the ability of laboratory culture conditions to fully support fertilized eggs with all genetic backgrounds. Given that more than 100,000 cycles of IVF are performed in the United States alone each year, with more than 30,000 babies born, this is clearly a major women's and children's health issue in urgent need of research.

Women volunteering to donate eggs for stem cell research also go through COH, a topic of some debate among groups concerned that women will be exploited for the sake of deriving stem cells for the more wealthy members of society. A potential risk of COH is an unusual over-response by the ovary to the artificial gonadotropin stimulation that results in abnormally high levels of estrogen production, swelling of the entire ovary to several times normal size, and accumulation of fluid in the abdomen. This serious complication, known as ovarian hyperstimulation syndrome, is the reason that all women undergoing COH should be monitored closely for estrogen production in response to gonadotropin injections so they may be discontinued before hyperstimulation occurs.

Once the eggs are collected, they are exposed to sperm. Until the early 1990s, each egg was exposed to approximately 100,000 to 500,000 sperm for fertilization. Remarkably, the incidence of more than one sperm entering the egg was less than 10%, even in the presence of such an overwhelming number of sperm. This indicates eggs have a remarkable innate defense mechanism against polyspermy.

As described in Chapter 4, a method of physically transferring a single sperm into the cytoplasm of the egg, termed *intracytoplasmic sperm injection* (ICSI), was developed. This method not only markedly improved rates of fertilization by less vigorous sperm, it has allowed an opportunity to study human egg fertilization at a new level.

Molecular Biology of Oocyte Meiosis

Human eggs are arrested in their development for a minimum of one and as many as five decades. As such, they represent not only the largest but probably the most quiescent long-lasting human cell. Although this characteristic is reminiscent of stem cells, a principal difference is that human eggs do not undergo self-renewal, and they are arrested in late G2/early M phase of the cell cycle, the significance of which is not known with certainty.

Most eggs in the human ovary exist as primordial follicles. This term refers to a relatively small oocyte, approximately 35 microns in diameter, surrounded by fewer than a dozen granulosa cells, flattened around the oocyte. There is no zona pellucida. This novel cellular friendship functions to support viability and developmental arrest of both the oocyte and the granulosa cells. Several lines of evidence have revealed the existence of communication in both directions: The egg expresses specific gene products that signal the granulosa cells, and the cells communicate with the egg cytoplasm through cytoplasmic projections, especially once the primordial follicle is selected to enter the monthly egg maturation cycle.

As discussed in Chapter 5, genetic manipulations of mice have revealed the existence of a number of gene products that are expressed specifically by eggs (Table 3.1). An important class of genes are those that from the zona pellucida, once thought to be synthesized by the granulosa cells, now known to be produced by the egg.

Studies of mouse models have revealed the existence of both egg and ovarian factors that influence meiosis. Eggs removed from the ovaries of female mice after puberty spontaneously resume meiosis in laboratory culture, unless the culture conditions are adjusted to increase the intracellular concentration of the second messenger, cyclic AMP. Eggs removed from ovaries before puberty remain arrested at the germinal vesicle stage. Taken together, these observations suggest prepubertal eggs lack cellular factors necessary to resume meiosis, whereas postpubertal eggs

TABLE 3.1	Egg-Specific Gene Products	
Gene	**Product**	**Role**
Zp1, Zp2, Zp3	Zona pellucida proteins 1, 2, 3	Zona pellucida
Gdf-9	Growth differentiation factor 9	Follicle growth
FIGα	*Transcription factor*	Expression of Zp1,2,3 and formation of primordial follicle
osH1	Oocyte-specific histone 1	Egg and sperm chromatin
Mater	MATER, 134-kilodalton protein	Unknown
mWee1B	Oocyte-specific kinase	Maintains meiotic arrest
OAS1D	Oligoadenylate synthetase	Follicle development
Nobox	Homeobox protein, NOBOX	Oocyte-specific gene expression

have acquired the factors, and meiosis is inhibited by overriding ovarian factors that maintain elevated concentrations of cyclic AMP. The ovarian factors responsible for suppressing meiosis for several decades in the human are unknown.

As discussed in Chapter 2, early work suggesting that factors outside the nucleus, in the cell cytoplasm, controlled the cell cycle came from studies of frog eggs. Landmark publications in 1971 by Masui and Markert and Smith and Ecker revealed that transfer of cytoplasm from metaphase II frog eggs into germinal vesicle stage frog eggs initiated meiosis in the germinal vesicle stage eggs. They coined the term **maturation promoting factor** (**MPF**), which was eventually purified by Maller in 1988. Molecular characterization of frog egg MPF revealed it was composed of two cell cycle proteins, Cdc2 and cyclin B, described in Chapter 2. Studies that followed demonstrated the same pathways in mammalian eggs. Thus, cell cycle regulators are also factors important to the resumption of meiosis.

But unlike dividing cells, mammalian eggs, including human, undergo another arrest at metaphase II. In the same series of experiments that led to the discovery of MPF, Masui and Markert transferred cytoplasm from a metaphase II–arrested frog egg into cleaving frog embryos and observed an arrest of embryo cleavage at M phase of the cell cycle. This indicated that metaphase II arrest was also controlled by cytoplasmic factors, which the authors termed *cytostatic factor* (CSF).

Subsequent work in several laboratories (see sidebar) led to the identification of a kinase, cMos, as an important component of CSF. cMos was originally identified as a cancer-causing gene (vMos) in a virus (Moloney sarcoma virus) that leads to sarcomas in mice. The normal cellular counterpart, cMos, was found to be uniquely expressed in ovaries and testes. Its physiologic role in the testis remains to be elucidated, but its role in the ovary appears to be confined to the egg.

MPF: maturation (or mitosis) promoting factor, a dimer of Cdc2 (Cdk2) and cyclin B.

cMos Activity
Metaphase II Arrest

Two groups of cancer scientists, led by Debra Wolgemuth and George vandeWoude, respectively, reported an unusual finding in 1985. A cancer-causing gene, *vMos*, first identified in Moloney sarcoma virus, was expressed specifically in testes and ovaries. *vMos* was known to encode a protein kinase that led to death in the cells infected by the virus. Subsequent independent studies by two other investigative teams, led by Geoffrey Cooper and John Eppig, respectively, revealed that cMos was expressed by eggs and not granulosa cells and that it played an important role in maintaining oocyte arrest at metaphase II. Results from the Cooper group indicated that inhibition of cMos protein synthesis actually resulted in oocyte activation, as demonstrated by pronuclear formation and cell cleavage; the results from the Eppig laboratory indicated that inhibition of cMos resulted in the loss of metaphase chromosomes but did not lead to egg cleavage. The presence of active cMos was confirmed in human eggs by the Cooper group.

These observations suggested an important role for cMos in reproduction. To understand the role, two other teams of investigators generated mice engineered without the gene for cMos. Eggs collected from the cMos "knock-out" mice demonstrated the same ability to undergo spontaneous activation as originally observed by the Cooper team. Taken together, the data suggest either that the experiments in the Eppig laboratory failed to completely inhibit cMos synthesis, suggesting two possible roles for cMos, one maintaining metaphase arrest and a second blocking mitosis, or that successful egg cleavage once cMos is inactivated requires specific laboratory conditions. Importantly, however, the females with the cMos gene deleted were not sterile, although they were subfertile. Thus, cMos expression enhances reproductive success but is not absolutely essential. Given the importance of reproduction to species survival, the existence of multiple families of genes that enhance reproductive success seems highly likely.

In keeping with the predictions, the ovaries of the females with cMos "knocked out" routinely contain dermoid cysts resulting from spontaneous resumption of meiosis within the follicle.

As described in Chapter 2, MPF is a complex of cyclin B and Cdc2 (see Figure 2.14). Cyclin B is synthesized during S phase and complexes with Cdc2. Activation of the complex by Cdc25 leads to kinase activity (see Figure 2.14) with a broad spectrum of targets, including other enzymes and structural proteins. Importantly, one of MPF's targets for phosphorylation early in M phase is the degradation machinery for cyclin B, which normally occurs after the metaphase plate is fully formed. Thus, to arrest the egg at metaphase II, it is necessary to block the degradation of cyclin B. This is the function of cMos. It phosphorylates a kinase, mitogen-activated protein (MAP kinase, which plays a central role in maintaining chromosomes in metaphase after the extrusion of the first polar body. MPF activity, as measured by the phosphorylation of histone type 1 (described in Chapter 2), decreases slightly at the time the first set of chromosomes is extruded, then increases again to maintain the metaphase II chromosomes for up to a few days in human eggs. Therefore, a fundamentally important aspect of egg activation is the elimination of cMos activity. Although not yet identified, the factors that eliminate cMos activity and allow the destruction of cyclin B to occur must be stockpiled within the egg's cytoplasm, awaiting activation.

These profound cytoplasmic controls on the egg cell cycle serve to emphasize the importance of cytoplasmic stockpiles to the success of egg activation. Moreover, the egg-specific controls need to be entirely replaced by dividing cell cycle controls within a brief interval after egg activation to allow cell cleavages to proceed normally. As demonstrated by the original studies with frog egg cytoplasm, factors in metaphase II–arrested eggs can bring about the arrest of cleaving blastomeres. Studies in animal models, principally the mouse, have revealed a rapid degradation of the stockpiles of mRNAs in the egg beginning at the time of fertilization. Such degradation may be as essential to the successful transition from oocyte to embryonic control of gene expression as new RNA synthesis from the cleaving blastomeres. This area is in urgent need of additional research.

The advent of high-throughput gene sequencing methods has allowed the compilation of libraries of mRNAs expressed at each stage of development. Taken together, mRNAs from 9,718 genes were detected for stages ranging from unfertilized eggs to blastocysts. Surprisingly, half of the genes were novel, revealing that a considerable fraction of mouse genes are dedicated to egg and early embryo cell functions. A total of 291 mRNAs was cloned specifically from eggs and disappeared after egg activation. Importantly, fewer than 10% corresponded to known genes, demonstrating how much remains to be learned about egg biology. Examples of the mRNAs detected specifically in unfertilized eggs are listed in Table 3.2.

TABLE 3.2	Gene Expression Specific to Unfertilized Mouse Eggs
GenBank No.	**Gene Product**
M18252	Mouse intracisternal type A particle RNA
S78271	Mitosis-specific chromosome segregation protein
M22995	Human *ras*-related protein, Krev-1
D50264	Phosphatidylinositol glycan class F
U20159	pp76, tyrosine phosphoprotein ligand of Grb2
Y15740	cMos
J04022	Calcium-dependent ATPase
U95116	PAF acetylhydrolase Ib alpha
X60672	Actin-capping protein, radixin
M80631	G protein alpha 14
Y08460	Mdes, mouse spermatogenesis protein

Summary

The egg is a giant cell with highly specialized functions. It begins its life cycle in the ovary of the fetal female, but it is not fully mature until it is released from the ovary during the monthly hormone cycle of postpubertal females. The process is termed *ovulation*. The mature human egg is ovulated part way through meiosis, the unique cell cycle that results in a reduction of the number of chromosomes to half the normal number, resulting in one of each, instead of a pair of each. Under normal circumstances, only one egg is recruited for ovulation each month. For reasons not understood, many eggs are ovulated that are not sufficiently robust to achieve development of a normal baby. Advances in understanding the hormones that regulate ovulation led to the development of hormone therapies for infertility treatment and a new medical specialty, reproductive endocrinology. Reproductive endocrinologists are the practitioners of assisted reproductive technologies. A central tenet of assisted reproductive technologies is the hormone stimulation of several eggs, instead of a single egg, to mature to increase the possibility that one has the capacity for full development of offspring.

At ovulation, the human egg is poised to initiate the transformation of itself and a sperm into a totipotent two-cell embryo. To accomplish this, the egg's life cycle includes the accumulation of specific gene products, some of which are put in place during the last few days before ovulation. The egg's gene products must be deployed in a specific sequence and immediately bring about several cycles of faithful replication of DNA and cell division. The molecular processes responsible for the egg's extraordinary role in life are poorly understood. New technologies for studies of genes specifically expressed by this remarkable cell have begun to reveal the complexity and unique nature of this early stage of development. The additional need for eggs to help create pluripotent stem cells, instead of babies, is expected to provide additional impetus to this critically important science.

Review Questions (True or False)

1. The human egg is 250 times the size of most adult cells.
2. Meiosis is the process whereby only one chromosome of each pair is retained.
3. During egg maturation, the chromosomes inherited from the mother and the father recombine into new chromosomes unique to that particular egg.
4. The final stages of egg maturation occur once a month under the control of protein hormones released by special glands in the brain.
5. The egg is assisted in its growth and accumulation of unique factors by surrounding cells, termed *granulosa cells*.
6. The factors inside the egg that are responsible for the initiation of fetal development have been known for several decades.

ADDITIONAL READINGS

Combelles, C. M. H., Carabatsos, M. J., Kumar, T. R., Matzuk, M. M., and Albertini, D. F. (2004). Hormonal control of somatic cell oocyte interactions during ovarian follicle development. *Mol Reprod Dev* 69: 347–355.

Dean, J. (2002). Oocyte-specific genes regulate follicle formation, fertility and early mouse development. *J Reprod Immunol* 53: 171–180.

Erickson, G. F. (2003). *Morphology and physiology of the ovary*. Endotext.com, www.endotext.org/female

Gilbert, S. F. (1999). *Developmental Biology,* 5th ed. Sunderland, MA: Sinauer Associates.

Hunt, P. A. (1998). The control of mammalian female meiosis: factors that influence chromosome segregation. *J Assist Reprod Gen* 15: 246–252.

Khan-Sabir, N., and Carr, B. R. (2005). *The normal menstrual cycle and the control of ovulation*. Endotext.com, www.endotext.org/female

Matzuk, M. M., Burns K. H., Viveiros, M. M., and Eppig, J. J. (2002). Intercellular communication in the mammalian ovary: oocytes carry the conversation. *Science* 296: 2178–2180.

Sagata, N. (1996). Meiotic metaphase arrest in animal oocytes: its mechanisms and biological significance. *Trends Cell Biol* 6: 22–28.

Serafica, M. D., Goto, T., and Trounson, A. O. (2005). Transcripts from a human primordial follicle cDNA library. *Hum Reprod* 20: 2074–2091.

Soyal, S. M., Amieh, A., and Dean, J. (2000). FIGa, a germ cell-specific transcription factor required for ovarian follicle formation. *Development* 127: 4645–4654.

Stanton, J. L., Bascand, L., Fisher, L., Quinn, M., Macgregor, A., and Green, D. P. L. (2002). Gene expression profiling of human GV oocytes: an analysis of a profile obtained by serial analysis of gene expression (SAGE). *J Reprod Immunol* 53: 193–201.

Tanaka, Y., Kato, S., Tanaka, M., Kuji, N., and Yoshimura, Y. (2003). Structure and expression of the human oocyte-specific histone H1 gene elucidated by direct RT-nested PCR of a single oocyte. *Biochem Biophys Res Commun* 304: 351–357.

CHAPTER 4

The Activated Egg

We shall not cease from exploration and the end of all our exploring will be to arrive where we started and know the place for the first time.

T. S. Eliot

OVERVIEW

The egg arrested in metaphase II and surrounded by thousands of cumulus cells is released at rupture of the ovarian follicle and swept into the fallopian tube like a giant bride with a million attendants. Stockpiled in the egg's cytoplasm are stores of messenger RNAs and proteins that must be released and activated in a precise cascade demanded by normal developmental processes.

As discussed in Chapter 2, dividing somatic cells systematically synthesize and then degrade the components of the cell cycle according to each transition phase. For example, the many proteins necessary to replicate all the DNA in the chromosomes are assembled at the G1 to S transition, and then most are degraded in G2 immediately after the checkpoint for DNA synthesis errors. Dismantling the DNA replication machinery ensures that DNA synthesis does not occur at other times in the cell cycle, thus maintaining accurate gene copy numbers. Similar strategies are in place for other cellular components whose actions are restricted to one phase of the cell cycle, such as the cyclins.

This is not the strategy adopted by the egg. Even though it is arrested at metaphase II, relatively high levels of DNA polymerase activity are detectable in mouse and rabbit eggs. There must be an evolutionary or reproductive advantage to having all components possibly needed for egg activation and early cleavages stored within the egg cytoplasm. The task for the egg therefore is to deploy its stockpiles in the systematic and precise manner needed for faithful

DNA replication and cell division. Moreover, it is important to remember that eggs have two unique cell cycle arrests: one in M phase when the egg is at the germinal vesicle stage and arrested in the prophase of meiosis for up to four or five decades and the other at the time of ovulation when the egg is arrested at metaphase II (discussed in Chapter 3).

The cell cycle arrest of the germinal vesicle stage egg is highly unusual because it involves a lengthy nondividing state usually reserved for G0 (discussed in Chapter 2), but the giant cell is actually arrested in early M phase during much of its growth cycle. Germinal vesicle stage eggs from several species, including cow, sheep, mouse, and human, collected directly from growing ovarian follicles and placed in culture in the laboratory can spontaneously undergo migration and breakdown of the germinal vesicle and resume meiosis. It is important to note that although most eggs that spontaneously resume meiosis in culture can undergo activation, including fertilization, they frequently fail after one or two cleavage divisions. This indicates that the capacity to resume meiosis precedes full maturation of the egg's potential to support development. The capacity to support fertilization follows, but because many fertilized eggs arrest at early stages, the capacity for full development must require additional egg factors. This is an important point, because developmental biology literature is replete with references to egg maturation, which is used to describe simply the capacity to resume meiosis. The meaning would be clearer if the term *egg maturation* were reserved for those eggs with full developmental potential.

Therefore, factors within the ovary must actively maintain the germinal vesicle stage of the egg to allow full maturation of its cytoplasm and accumulation of stockpiles necessary not only for early cell cycles, but also for healthy embryonic and fetal development. The identity of all the factors is not known, but the intracellular messenger, **cyclic adenosine monophosphate (cAMP)**, plays an important role. Maintaining high concentrations of cAMP during culture of germinal vesicle stage eggs blocks the breakdown of the germinal vesicle and the resumption of meiosis. A full understanding of the factors that direct the egg's unusually long arrest in early M phase could lead to new approaches for deriving stem cells for therapy.

As described previously, germinal vesicle stage eggs can be activated to resume meiosis simply by removing them from the ovary. In nature, as discussed in Chapter 3, the luteinizing hormone (LH) surge leads to the resumption of meiosis while the egg is still in the ovary. Therefore, it is assumed that the LH surge overcomes the ovarian blockade to meiosis, perhaps by stimulating the production

cAMP: a common abbreviation for cyclic adenosine monophosphate, a molecule formed in response to hormonal stimulation of cells; it is classified as a second messenger because it travels through the cell and activates kinases.

of progesterone. This is an ongoing area of research. The egg is programmed to locate the metaphase I chromosomes near the periphery, close to the plasma membrane of the egg. This leads to unequal division of egg cytoplasm when the polar body extrudes with half the chromosomes but only a tiny fraction of the cytoplasm. The remaining cytoplasmic stores are thus available to the activated egg.

The completion of meiosis I and the extrusion of the first polar body lead to chromosome rearrangement and the second arrest point for the egg at metaphase II. As discussed in Chapter 3, cMos is the kinase whose activity is essential for maintaining arrest at metaphase II. Because arrest at metaphase II is unique to eggs, it is not surprising that cMos is specifically expressed in eggs and not in other cells in the ovary. Moreover, it follows that cMos is not expressed in cycling somatic cells outside the ovary because metaphase arrests are rarely observed. Thus, cMos is an example of proteins whose expression is peculiar to eggs; they are part of the group of genes expressed in eggs collectively referred to as **maternal messages**. A full understanding of maternal messages and the controls on their deployment could lead to valuable new approaches for the derivation of stem cells from activated eggs.

Eggs undergo activation in a surprising variety of ways, described in this chapter as spontaneous egg activation, artificial egg activation, egg activation by sperm, and egg activation after nuclear transplantation.

Spontaneous Egg Activation

Because cMos plays such an essential role in maintaining metaphase II arrest, it follows that defects in cMos expression or activity could lead to spontaneous egg activation. Evidence that spontaneous activation actually occurs in nature stems from the formation of **dermoid cysts**, a relatively common occurrence in women. Available evidence suggests that failure of the follicle to rupture and release the egg at metaphase II can result in egg activation within the follicle. This gives rise to the cluster of cells characteristic of dermoid cysts. Further development can lead to **teratomas**, benign growths that include collections of differentiated cells that exhibit some characteristics of adult tissues, such as muscle and bone. As described in Chapter 3, mice genetically engineered without the *cMos* gene developed ovarian teratomas at a high rate.

This type of egg activation obviously does not involve fertilization by sperm and is termed *parthenogenesis,* from the Greek words parthenos, meaning virgin, and genesis, meaning beginning. This naturally occurring phenomenon emphasizes that sperm are not required for egg activation, an observation further supported by many studies of parthenogenic activation of eggs of most species. Eggs activated without sperm are termed **parthenotes**. The observation that dermoid cysts and

Parthenote: a cleaving egg that is activated without being fertilized by sperm.

teratomas contain several cell types supports the concept that parthenogenic activation of human eggs may lead to the derivation of pluripotent human stem cells, as has recently been shown for primate eggs (see Chapter 1). For this to occur, several modifications of normal fertilized egg processes are necessary, as discussed in the following section on artificial egg activation.

Release of stored calcium ions in precise pulses, like a metronome pacing the music, plays a major role in egg activation. The calcium ions serve as **second messengers** within the egg cytoplasm and in some way bring about the deployment of other stored components. The first wave participates in the completion of meiosis II, and the waves that follow initiate cell cleavages (Figure 4.1). Although many of the details are still unknown, it is clear that the calcium pulses must occur in carefully timed intervals for a period of several hours to ensure sustained activation of the egg. The periodicity is species specific. Rodent and human eggs release calcium at peak intervals of 2 to 4 minutes, whereas cow and pig eggs release calcium at intervals of 15 to 30 minutes. It is as if each pulse is responsible for activation of components whose activities must follow previous events and carry out functions essential to subsequent events.

One early event in all species studied is the degradation of cMos, which allows meiosis to resume. Although evidence is limited, as suggested previously,

Second messenger: a molecule created in response to a specific stimulus, such as hormone binding to a cell membrane receptor, which in turn brings about a specific cellular response such as activation of a kinase or a phosphatase.

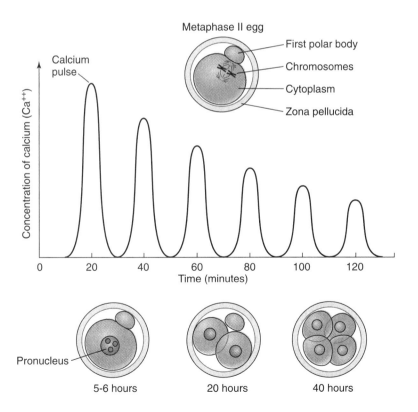

FIGURE 4.1
Release of calcium ions stimulates egg activation.

FIGURE 4.2
Haploid and diploid egg parthenotes.

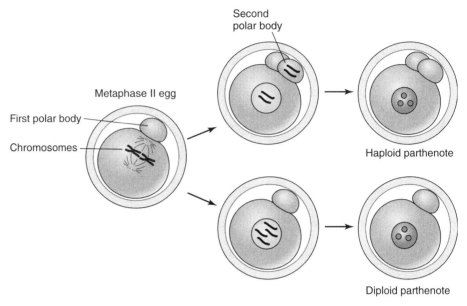

eggs deficient in cMos may be more readily activated and less dependent on the strict regimen of calcium peaks. Under the usual rules of meiosis, half the metaphase II chromosomes are extruded as a second polar body and the other half are enclosed in a nuclear membrane within the egg (Figure 4.2). This creates a haploid nucleus within the egg.

Alternatively, if formation of the second polar body is prevented, then all the metaphase II chromosomes become enclosed within the egg's nucleus and it becomes a diploid cell. Interestingly, a landmark paper written by John Eppig in 1977 describing ovarian teratomas in mice and humans reported genetic evidence that teratoma formation did occur after extrusion of the first polar body and that the resulting tumors were diploid, suggesting either that extrusion of the second polar body was suppressed or that the haploid set of chromosomes underwent DNA replication before activation and cell division was initiated, resulting in the cluster of cells within the ovarian follicle. If the cells initiate regular cell cycles, then they may also respond to growth factors and differentiate accordingly, as described in Chapter 11. This could explain the formation of teratomas with several types of differentiated cells.

Artificial Egg Activation

One form of egg activation with the potential to give rise to two equal cells is to reprogram the germinal vesicle stage egg to bring about an equal cell division, yielding two "half eggs" instead of one polar body and one egg (Figure 4.3). This experimental possibility has received little attention by developmental biologists

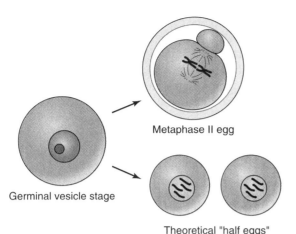

Germinal vesicle stage

Metaphase II egg

Theoretical "half eggs"

FIGURE 4.3
Theoretical splitting of the egg at
meiosis I.

but could theoretically double the number of "eggs" available for stem cell production. Each half egg would have a normal diploid number of chromosomes and theoretically could be activated to continue cell division parthenogenically. This possibility is an example of an area of research needed to improve the resources for stem cell derivation.

Development after parthenogenetic activation of eggs arrested at metaphase II, most frequently frog, mouse, and rabbit egg model systems, has been studied by numerous investigators for several decades. Egg activation without fertilization is regarded as a useful experimental tool with which to probe and understand those early developmental functions that are dependent on the molecular legacy of the egg. It also provides an opportunity to derive individuals that would be homozygous for maternal genetics, as discussed in Chapter 5.

Careful electrical pulses and several types of chemicals, including molecules as simple as ethanol, can bring about activation of metaphase II eggs in culture. Egg activation and one initial cleavage happens more readily in "aged" eggs than in freshly ovulated eggs, suggesting that cMos activity decays over time and cannot be renewed indefinitely. Aged eggs do not successfully sustain cleavage, however, even when activated by fertilization with sperm, suggesting that other maternal stores of essential cell cycle components may also decay.

Chemical activation includes compounds that inhibit protein synthesis, block microtubule assembly, stimulate release of calcium, and interact with egg plasma membrane receptors. Each species responds uniquely to such stimuli, supporting the concept that although mammalian eggs share many features in common, details of developmental activation are species specific. Thus, the precise activation regimen that will lead to sustained cleavage of human eggs is an active area of research. The goal is to suppress the extrusion of the second polar body, thus retaining both sets of egg chromosomes that become the content of a single large pronucleus.

If the egg can undergo activation, cleavage, and differentiation of different cell types, the possibility exists that under the right circumstances, offspring could form. There is no evidence for this in humans or other mammals, but this is a common occurrence among some insects, such as aphids. It is not known at this time whether mammalian eggs can be reprogrammed for full development of an offspring or whether the requirement for a placenta limits developmental potential. It is clear that some genes are expressed differently if they are inherited from the father rather than from the mother. This suggests that the environment of the ovary and the testis influence specific genes differently, a process termed *imprinting*. Several characteristics of zygotes that relate to this question are described in Chapter 5.

Imprinting: a modification to a gene, without altering the base sequence, that occurs in the ovary or in the testis and influences what cells can express the gene.

Egg Activation by Sperm

Precisely how sperm contact activates the egg is an ongoing area of research. Animal models must be used principally in the United States because the U.S. Congress has decreed that although research on fertilized human eggs "is meritorious and should be done for the good of society and humanity, the research will not be funded by taxpayer dollars." Therefore, the limited research that is done on fertilized human eggs in this country is done with private funds. The first reported attempts to fertilize human eggs in laboratory culture took place in Brookline, Massachusetts in the late 1930s and early 1940s (see sidebar).

Sperm entry into the egg leads to a cascade of events. Calcium spikes begin at regular intervals, thought to be mediated by factors brought into the egg by the sperm itself, described in more detail in the molecular biology section of this chapter. Such sperm factors have only recently been described and constitute an active area of research. The most striking distinction between egg activation after sperm penetration and artificial egg activation is the nature of the calcium spikes. Sperm penetration produces well-defined calcium spikes at precise intervals, sustained for the exact period of time necessary to deploy egg stockpiles of proteins and other molecules needed to support early embryonic events.

Meiosis resumes at sperm entry, and M phase is completed with the extrusion of the second polar body. Simultaneously, the sperm head is rapidly remodeled into a much larger round nucleus by components in the egg cytoplasm (Figure 4.4). Sperm DNA is packaged tightly into the sperm head nucleus with unusual proteins, spermidine and protamine, which must be removed and replaced with standard somatic cell chromosome proteins, the histones. The remodeled sperm chromosomes are enclosed within a nuclear membrane, and DNA replication begins almost immediately. Sperm do not contain the enzymes for DNA replication, so the DNA replication complex must be garnered from the egg cytoplasm. The sperm head remodeling process takes approximately 6 to 8 hours and is analogous to G1 in a somatic cell cycle (see Figure 2.8). While sperm head remodeling is occurring, the remaining egg chromosomes are enclosed within a second pronucleus, and replication of egg chromosomes also begins (Figure 4.4).

Because the sperm and egg chromosomes are individually haploid, the newly formed sperm nuclei are termed *pronuclei*. With a diameter of approximately

Pronucleus: the large nucleus with a few large prominent nucleoli formed around the chromosomes of an activated egg and the sperm head, if fertilization occurred.

Activation of Human Eggs
In Vitro Fertilization and Cleavage of Human Ovarian Eggs

Research that began in the late 1930s was partially reported in *Science* (1944; 100: 105–107). The goal was to observe to better understand fertilization and cleavage of human eggs. At the time the work was performed, fertilization of eggs by sperm, the fundamentally most important process for all species that sexually reproduce, had been observed in only a very few species. Dr. John Rock was an obstetrician/gynecologist at Harvard Medical School. His duties included caring for women at the Fertility Clinic Laboratory, Free Hospital for Women, in nearby Brookline. With the aid of his laboratory assistant, Ruth Menkin, they had recovered approximately 800 eggs from ovarian follicles of ovaries removed at hysterectomy. Of those, 138 were observed after exposure to sperm, and the 1944 report described the eggs from three women who underwent cleavage.

Mrs. DD: One cumulus-enclosed egg, recovered from a 2.3-centimeter follicle, was cultured in Mrs. DD's serum for 27 hours and then exposed to washed sperm for 1 hour in Locke's solution, during which time it was continuously viewed. "The spermatozoa showed great activity throughout the period of observation; they were clearly seen to travel through the interstices of the loose cellular formation surrounding the egg, and many were noted in active motion just outside the ovular boundary. At the end of one hour, the ovum was transferred to fresh serum from a post-menopausal patient." Forty hours later, it had cleaved to two blastomeres, each measuring 86 microns in diameter.

Mrs. RP: Thirteen eggs were recovered from developing follicles, cultured for 22 hours as before, exposed to sperm for 2 hours, transferred to postmenopausal serum, and observed 45 hours later. One (100–113 microns) had cleaved to two cells and had numerous sperm heads attached to the zona pellucida. The blastomeres measured 88–58 and 58–105 microns, respectively.

Mrs. JD: Four eggs were recovered from ovaries removed to seal the fimbriae because of "tuberculous endometritis." The eggs were cultured and observed as described previously. Two cleaved, one to three blastomeres. The egg measured 103–127 microns, the largest blastomere 97–73 microns and the two smaller ones were 62–62 and 50–63 microns, respectively.

These results, with respect to timing of cleavage, were in general agreement with the observations made on cleaving monkey eggs in 1933 by Lewis and Hartman (Carnegie Institute of Washington, Publication 443, *Contributions to Embryology* 1933; 24: 187). In all, 4 of the 18 eggs recovered from the ovaries of the three women exhibited one or two cleavages. Did Menkin and Rock observe the first fertilized human eggs or the first human parthenotes?

40 microns, the pronuclei are also extraordinarily large, leading to more open scaffolding of gene sequences than is present in somatic cells. The pronuclei are characterized by a few large nucleoli and thus resemble chocolate chip cookies (Figure 4.4). The two pronuclei are unique to the zygotic stage, as discussed in Chapter 5.

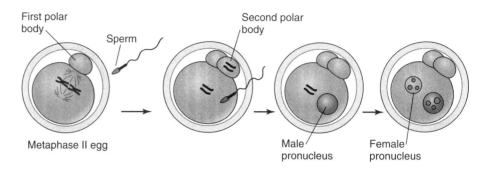

FIGURE 4.4
Sperm penetration and formation of pronuclei.

First polar body

Sperm

Second polar body

Metaphase II egg

Male pronucleus

Female pronucleus

Egg Activation After Nuclear Transplantation

Given the ability of an activated egg to remodel sperm heads, it seemed likely that it could remodel somatic cell nuclei. Early interest in this process (described in more detail in Chapters 7 and 10) was for the purpose of testing the developmental potential of adult nuclei from frog somatic cells. After all, if all cells contain all genes, then it seemed theoretically possible that any cell in the body could direct the development of a new individual. Nearly 50 years ago, two groups of scientists reported the experimental results of nuclear transplantation into activated frog eggs. Nuclei were transferred into the egg cytoplasm by a variety of methods (see Chapter 7). They discovered significant variations in the potential of nuclei to become remodeled by the activated egg. The more advanced the developmental stage of the transplanted nucleus, the less successful it was at supporting the development of a new embryo. Nonetheless, these experiments proved the principle that cells differentiated to become constituents of specific tissues and organs have retained all their genetic information.

Most nuclear transplant experiments involve removal of the egg's metaphase II chromosomes that have come to lie adjacent to one side of the egg plasma membrane, in preparation for extrusion of the second polar body. Once the chromosomes have been removed, the egg is sometimes termed a *cytoplast* because it is no longer a cell on its own but a large bag of egg components poised to initiate development.

To direct cell division, a new set of chromosomes is transplanted into the egg. *Cloning* is an accurate term for this process because chromosomes from a single cell are placed within the egg cytoplasm, generally in the form of an intact nucleus from a somatic cell, such as a skin cell. As described in Chapter 7, this is sometimes accomplished by physical transfer of a nucleus with the aid of a pointed micropipet that penetrates the egg's membrane and other times is accomplished by fusion of a somatic cell with the egg's plasma membrane. The result is a reconstructed egg with a new set of chromosomes and some cellular organelles, such as mitochondria.

The somatic cell nucleus is remodeled into one large pronucleus that resembles the single pronucleus in the parthenote egg. Successful nuclear remodeling yields chromosomes that have reverted to a fully undifferentiated state, thus poised to guide development of the cleaving egg into pluripotent stem cells, capable of both self-renewal and development into many cell types. Under the right circumstances, including formation of an embryo and fetus if transferred into a recipient uterus, as described in Chapters 7 and 10, this technology is also the basis for animal cloning research.

Molecular Biology of Egg Activation

Calcium ion signaling is a common cellular tool for responding to a variety of stimuli, including the need to alter cell physiology or to grow and divide. The development of molecules that fluoresce when bound to calcium provides cellular probes for calcium release that can be easily tracked by fluorescent microscopes, thus paving the way for thousands of reports of patterns of calcium ion release. Eggs are no exception.

Occasional spontaneous peaks of calcium ion release are detectable once the egg attains the maturity to resume meiosis. Calcium ions are stored in intracellular protein networks termed *reticulum*, and their release is brought about by the activation of calcium channels. Once released, calcium ions serve as activating cofactors for many enzymes, including kinases, such as protein kinase C. The pulsatile nature of the system provides an ingenious mechanism for cyclic, rather than sustained, activities, essential elements of DNA replication and cell division.

Important calcium release gates are the receptors for a relatively small freely soluble modified sugar, **inositol triphosphate** (IP_3) (Figure 4.5). IP_3 is a phosphorylated hexose bound by a phosphate ester to plasma membrane lipids that project into the cytoplasm. It is cleaved by a **phospholipase C**, freeing it to migrate through the cell to the endoplasmic reticulum where it encounters its receptor, and once bound, calcium ion is released. Molecules that bind to receptors are termed *ligands*.

Hence, an important initial step in calcium release is activation of the enzyme, phospholipase C. Activation of phospholipase C is brought about by the action of another receptor response element, called a **trimeric G protein**. G proteins are a large family of response elements that are coupled to a wide variety of receptors, thus linking calcium release to a wide variety of extracellular stimuli. There are several isoforms of phospholipase C, most of which appear to be present in eggs.

In the egg, the release of calcium ion itself triggers the activation of phospholipase C and the release of IP_3, thus propagating the intermittent calcium pulses. At the same time, the other cleavage product of phospholipase C, **diacylglycerol**, diffuses through the cytoplasm and, in combination with calcium ion, activates protein kinase C, thought to play an important role in the resumption of meiosis. There are also several forms of protein kinase C, most of which are thought to be present and active in eggs.

FIGURE 4.5

Meiosis

As discussed previously, one approach to doubling the number of eggs potentially available for stem cell derivation is to manipulate the cleavage at meiosis I into two equal cells rather than the unequal division into a polar body and an egg arrested at metaphase II. Although this area has received little attention by developmental biologists, work to date indicates that an actin-rich region in the egg plasma membrane attracts and holds the first metaphase plate and that if the actin filaments are disrupted or dispersed, the metaphase plate migrates back to the center of the egg. Therefore, it seems theoretically possible to engineer equal cleavage divisions of the germinal vesicle stage egg, giving rise to two half eggs for progression to metaphase II (Figure 4.3). There would likely be no reproductive advantage to the creation of half eggs because of the limited developmental potential of the reduced cytoplasm, but there could be significant advantage to the creation of stem cells. The half eggs could be activated either by parthenogenesis or after nuclear transplantation. This requires suppression of the pathways involved in maintaining metaphase arrest.

Stimulating metaphase II–arrested eggs to resume development is first dependent on decreasing maturation promoting factor (MPF) activity, presumably by eliminating the actions of cMos. Because inhibitors of protein synthesis, such as cycloheximide, bring about activation, it is presumed that cMos is continuously translated during metaphase II arrest. Therefore, it is postulated that the first wave of calcium spikes brings about the degradation of cMos and the dephosphorylation of mitogen-activated protein (MAP) kinase, leading to the resumption of meiosis (Figure 4.4).

Zygote formation

As discussed in detail in Chapter 5, parthenote zygotes and nuclear transplant zygotes contain only one large pronucleus, whereas fertilized eggs contain two pronuclei. Parthenote zygotes will contain a normal set of chromosomes if extrusion of the second polar body is suppressed during the resumption of meiosis. Chemicals that inhibit protein synthesis, microtubules, or protein kinase activity are effective in blocking extrusion of the second polar body.

An example of an activation protocol used with mammalian eggs, including human, is a combination of a chemical termed an *ionophore* and a protein kinase inhibitor. An example of an ionophore is A23187, a molecule that can transport a charged metal ion, such as calcium, into a lipid-rich environment, such as a cell membrane. Adding A23187 to culture medium facilitates the uptake of calcium into the egg, thus artificially creating a calcium spike within the egg. Ionophore exposure is generally brief, less than 10 minutes.

An example of a protein kinase inhibitor is **6-dimethylaminopurine** (DMAP), reported to prevent extrusion of the second polar body and promote formation of pronuclei. Brief exposure to ionophore followed by a few hours of exposure to DMAP leads to activation of most mammalian eggs, including cow, goat, mouse, monkey, and human eggs. The efficiency of artificial activation is less than with fertilization, however, suggesting the sperm factor that initiates pulses of calcium release is far more precise than the chemical substitutes.

Electrical stimuli can also induce calcium spikes, as demonstrated by an elegant series of studies with rabbit eggs. Carefully timed periodicity of small electrical pulses resulted in higher efficiency of parthenogenic development of the rabbit eggs than the combination of chemicals described above.

Fertilization

Sperm entry into the egg releases a protein that plays a key role in egg activation by mobilizing calcium. Likened to "love's first kiss," sperm factor was originally described in 1996 and identified as a pentose phosphatase, but this is now known not to be its identity as a result of the work of several other laboratories. The strongest candidate at this time is a phospholipase C, termed *PLC zeta*, that has been isolated from several mammalian sperm species, including human. A purified form of PLC zeta efficiently creates calcium spikes characteristic of fertilization when injected directly into eggs. Realization that calcium stores are compartmentalized within cells, including egg endoplasmic reticulum, suggests calcium mobilization needs to be deployed in a specific fashion for normal development. Recruitment of extracellular calcium into the egg by ionophores is an interesting experimental model but is clearly not a comparable replacement for sperm factors. Derivation of stem cells from eggs activated without sperm exposure requires much additional research in this area.

Summary

At ovulation, the human egg is arrested midway through meiosis. Its chromosomes are poised near the periphery of its cytoplasm, ready to expel one member of each pair in the second polar body. Meiotic arrest is maintained by egg-specific kinases, including cMos, which must be inhibited to bring about egg activation. A central element of egg activation is pulsatile release of calcium ions.

Egg activation may occur spontaneously in the ovary, after exposure to certain chemicals and/or electrical activation, or at entry of a sperm. Spontaneous or chemical/electrical egg activation is termed *parthenogenesis* and may be accompanied by second polar body extrusion, leading to a haploid parthenote, or by retention of both members of the chromosome pairs, leading to a diploid parthenote. Fertilization by sperm stimulates second polar body extrusion to accommodate the chromosomes introduced by the sperm itself.

Nuclear transplantation involves first removing the egg's chromosomes, followed by insertion of a new diploid set of chromosomes, usually in the form of a somatic cell or its nucleus. The reconstructed egg is then artificially activated.

The first nucleus formed after all methods of egg activation is huge and is termed a *pronucleus*. The precise nature of the egg components that implement activation and chromosome remodeling to accommodate the formation of the large pronucleus is unknown. Advances in detecting gene expression in single cells have revealed some of the pathways invoked by the calcium ion pulses, as well as some of the genes specifically expressed in activated eggs. As of this writing, pluripotent stem cells arise from eggs deployed to remodel chromosomes to express genes essential to pluripotency, the capacity to continually divide and retain the potential to differentiate into all tissues. Changes to the DNA in chromosomes that influence gene expression include imprinting. Fully understanding the egg pathways that invoke pluripotentiality may lead to the ability to reprogram chromosomes without the requirement for eggs.

Review Questions (True or False)

1. Ovulated human eggs have extruded one polar body that contains a complete set of chromosomes.
2. The remaining egg chromosomes have rearranged with one member of each chromosome pair positioned opposite each other in an arrangement termed a *metaphase plate*.
3. Egg activation requires proteins from the sperm head.
4. Eggs can be activated by electricity or by chemicals that stimulate the initiation of pulses of calcium ions within the egg cytoplasm.
5. Eggs activated with their own chromosomes only are termed *parthenotes*.
6. The calcium ion pulses recruit stores of maternal factors that bring about the formation of the pronucleus and initiate DNA replication.
7. Imprinting is the process of adding genes from another individual to an activated egg.

8. Trimeric G proteins are important elements within the cell cytoplasm that respond to external stimuli.

ADDITIONAL READINGS

Balakier, H., Dziak, E., Sojecti, A., Librach, C., Michalak, M., and Opas, M. (2002). Calcium-binding proteins and calcium-release channels in human maturing oocytes, pronuclear zygotes and early preimplantation embryos. *Hum Reprod* 17: 2938–2947.

Ben-Aharon, I., Ben-Yosef, D., Amit, A., and Shalgi, R. (2005). Expression and immunolocalization of the calpain-calpastatin system in the human oocyte. *Fertil Steril* 83: 1807–1813.

Ben-Yosef, D., and Shalgi, R. (2001). Oocyte activation: lessons from human infertility. *Trends Mol Med* 7: 163–169.

Cibelli, J. B., Kiessling, A. A., Cunniff, K., Richards, C., Lanza, R. P., and West, M. D. (2001) Somatic cell nuclear transfer in humans: pronuclear and early embryonic development. *J Regen Med* 2: 25–31.

Ducibella, T., Huneau, D., Angelichio, E., Xu, Z., Schultz, R. M., Kopf, G. S., Fissore, R., Madoux, S., and Ozil, J. P. (2002). Egg-to-embryo transition is driven by differential responses to Ca(2+) oscillation number. *Dev Biol* 250: 280–291.

Goud, P. T., Goud, A. P., Leybaert, L., Oostveldt, P. V., Mikoshiba, K., Diamond, M. P., and Dhont, M. (2002). Inositol 1,4,5-trisphosphate receptor function in human oocytes: calcium responses and oocyte activation-related phenomena induced by photolytic release of InsP3 are blocked by a specific antibody to the type I receptor. *Mol Hum Reprod* 8: 912–918.

Kiessling, A. A. (2005). Eggs alone. *Nature* 434: 145.

Kurokawa, M., Sato, K., and Fissore, R. A. (2004). Mammalian fertilization: from sperm factor to phospholipase C zeta. *Biol Cell* 96: 37–45.

Lan, G. C., Han, D., Wu, Y.-G., Han, Z.-B., Ma, S.-F., Liu, X.-Y., Chang, C.-L., and Tan, J.-H. (2005). Effects of duration, concentration and timing of ionomycin and 6-dimethylaminopurine treatment on activation of goat oocytes. *Mol Reprod Dev* 71: 380–388.

Ozil, J. P. (1990). The parthenogenetic development of rabbit oocytes after repetitive pulsatile electrical stimulation. *Development* 109: 117–127.

Rosenblum, I. Y., and Heyner, S. (1989). *Growth Factors in Mammalian Development*. Boca Raton, FL: CRC Press.

Tesarik, J. (2002). Calcium signaling in human oocytes and embryos: two-store model revival. *Hum Reprod* 17: 2948–2949.

Van Blerkom, J. (1994). *The Biological Basis of Early Reproductive Failure in the Human: Applications to Medically Assisted Conception*. Oxford: Oxford University Press.

CHAPTER 5

The Zygote

Make everything as simple as possible, but not simpler.

Albert Einstein

OVERVIEW

The Greek *zygous* means "yoked" and the French *zygous* means "joined," terms that bring to mind teams of oxen yoked together to share the workload. For centuries, the word has been applied to the most unique and unusual developmental stage of all individuals, the zygote. In the zygote, for the first time, genetic information from each parent is enclosed within separate, distinct, abnormally large nuclei, **pronuclei,** which form after fertilization. Toward the end of the zygote stage, the pronuclei come to sit adjacent to each other within the egg cytoplasm and together assume the tasks of development (Figure 5.1). This chapter reviews the zygote, including the requirements for both the maternal and paternal genes contained within the pronuclei to give rise to a new individual. In this way, the pronuclei are truly the working team the term suggests.

The Pronuclear Stage Egg

The human egg, like the eggs of most mammals, completes meiosis when metaphase II arrest is overcome by sperm entry. The metaphase II chromosomes resume meiosis almost immediately when the sperm enters the egg or the egg is otherwise activated, and the second polar body is extruded. Thus, unlike sperm, human eggs do not naturally exist in a haploid state, and fertilization actually overlaps meiosis. Unlike the first polar body, the second polar body is haploid. The fate of the polar bodies is not known with certainty. The first polar body sometimes divides, sometimes disintegrates. There is no evidence that the second polar body divides; it is thought to disintegrate. There is renewed interest in polar bodies and their fate because of the possibility that analyzing their genetic composition may provide valuable clues about the

FIGURE 5.1
Human zygote.

genetics of the egg. Some couples undergo in vitro fertilization so their embryos may be analyzed for specific life-threatening disorders, such as Tay-Sachs disease or sickle cell anemia, known to be present in their genetic heritage. To avoid fertilizing genetically affected eggs, it is theoretically possible to analyze the polar bodies and in this way surmise the genetics of the egg.

The process of sperm head remodeling begins immediately on sperm entry and lasts approximately 6 to 8 hours. The exact processes involved in sperm head remodeling are not known with certainty and are an active area of research. Sperm chromosomes are tightly packed within the sperm head with sperm-specific proteins such as **protamine and spermidine**. Mature sperm are not transcriptionally active, thought to be related to the tight packing of chromosomes. Sperm chromosomes are decondensed within the cytoplasm of the egg, the sperm proteins replaced with histones, and the sperm chromosomes wrapped within a new nuclear membrane, giving rise to the male pronucleus.

While sperm head remodeling begins, the remaining egg chromosomes from meiosis II decondense into chromatin and become enclosed within a second pronucleus. As discussed in Chapter 4, this period is analogous to G1 in a somatic cell cycle (see Figure 2.8), except that the cellular machinery needed for DNA replication preexists in the egg cytoplasm.

The pronuclei have a diameter of approximately 40 microns, and each contains the haploid complement of sperm and egg genes. The reason for the extraordinarily large size of the pronucleus is not clear, but the circumstance leads to the possibility of more open scaffolding of chromosomes than is present in somatic cells. Such open scaffolding may ease constraints against gene expression, or DNA replication, as discussed at the end of this chapter. The pronuclei are characterized by a few large nucleoli, the organizing regions for ribosomal RNA, and resemble chocolate chip cookies (Figure 5.1). Pronuclei appear in human eggs approximately 6 to 8 hours after exposure to sperm and persist for approximately 10 to 12 hours, during which time DNA synthesis occurs within each separate pronucleus, in preparation for cell division. Interestingly, several studies in the mouse have shown that DNA replication in the male pronucleus actually begins before the initiation of DNA replication in the female pronucleus. This suggests that sperm head remodeling may involve components of the DNA replication complex. This possibility may be important to successful nuclear remodeling, as discussed in Chapter 7.

Although not studied directly in human eggs, metaphase II mouse, rabbit, and goat eggs contain DNA polymerase activity at much higher levels than single somatic cells in S phase. This is reminiscent of frog eggs, which contain sufficient DNA polymerase activity to replicate all frog DNA within minutes instead of hours. But S phase in mouse and rabbit zygotes does not appear to be shortened by the increased amount of DNA polymerase activity. It requires the standard 6 to 8 hours observed for most mammalian cells. Therefore, because the increased amount of DNA polymerase activity does not apparently lead to shorter time for DNA replication, it may simply reflect sufficient stores to support more than one replication cycle without new synthesis of the DNA replication complex.

The length of time it takes for DNA to replicate in fertilized human eggs is not known with precision, but available evidence suggests that it is comparable to the 6- to 8-hour DNA synthesis time it takes for human somatic cells. It is assumed, analogous to the mouse, rabbit, and goat, that unfertilized human eggs also contain extraordinary levels of DNA polymerase activity, as discussed in more detail in Chapter 6.

As described in Chapter 2, a fundamental feature of S phase is checkpoints to detect errors in DNA replication. The checkpoints allow the errors to be repaired before the cell cycle proceeds to cell division. Whether or not such a checkpoint exists in human zygotes is not known. The existence of such a checkpoint in somatic cells has been revealed by a variety of experiments, including purposefully damaging DNA during S phase (for example, by exposure to x-rays or to chemicals that damage DNA). Healthy cells postpone the transition from S phase to G2 until the DNA damage is repaired and then proceed. Some studies also reveal increases in repair enzymes in response to DNA damage.

But zygotes rarely arrest in S phase, suggesting either that DNA repair enzymes are abundant and can repair damage quickly or that there is no S phase checkpoint for DNA damage in zygotes. Because no single process is more fundamentally important to the development of offspring than faithful DNA replication in the zygote, such a checkpoint seems critically important. Nonrepaired errors in a DNA sequence would lead to damaged chromosomes in every embryonic cell. Clearly, defects in DNA replication and repair in zygotes could account for many of the failed embryos and is an area in urgent need of additional research.

Near the end of S phase, the pronuclei migrate to the center of the egg. The G2 phase is short, and M phase begins nearly concomitantly with the touching of the pronuclear membranes (Figure 5.2). Metaphase and anaphase within human eggs is less well studied than in animal species because of the U.S.

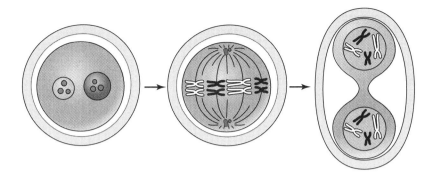

FIGURE 5.2
The first cleavage.

congressional moratorium against funding the research with taxpayer dollars. However, available evidence suggests that chromosome line-up and equal division into two daughter cells occurs in a fashion analogous to other cells. Cleavage proceeds immediately, delivering one chromatid from each parental chromosome into each daughter two cell. Because DNA replication is semiconservative, as discussed in Chapter 2, one DNA strand of each chromosome is the original strand from the sperm or egg and the other strand is the one newly replicated in the zygote. The two-cell stage thus establishes for the first time a new genetic makeup within each nucleus.

In the mouse egg, a bridge between the two new daughter cells persists for a few hours after the cleavage furrow forms, allowing cytoplasmic contents to be shared between the two cells for a few hours. Unlike somatic cells, the first egg cleavage is not dependent on expression of genes in either the male or female pronucleus, although that is not known with certainty for human eggs. Nor is it known if the cell cycle machinery recruited for the first S, G2, and M phases is degraded after the first cleavage, as in somatic cells. What is known is that the G1 phase of the two-cell egg is short, suggesting that the egg's DNA replication machinery persists into at least the second S phase, as discussed in Chapter 6. This is consistent with direct measurements of DNA polymerase activity in mouse and rabbit blastomeres, which revealed relatively constant levels of enzyme activity during early cleavages.

Manipulation of the Zygote

The clearly visible pronuclei provide the opportunity for several types of manipulations of the zygote to explore fundamentally important questions about developmental potential. Probably the first zygotic manipulations were performed by Hans Spemann in the early 1900s. As described in more detail in Chapter 7, Spemann was able to manipulate the zygotic nuclei of a fertilized newt egg into one half of an egg that developed normally. The other half of the egg failed to undergo any development, demonstrating that the information guiding embryonic development resided in the nuclei.

Other types of manipulations of mouse zygotes revealed the importance to embryonic development of the presence of both the paternal and maternal genes. The pronucleus developed from the sperm head is larger and forms earlier than the female pronucleus. Several teams of investigators have removed one pronucleus from a zygote and replaced it with a pronucleus from another zygote. In this way, it is possible to reconstruct zygotes with two female pronuclei and/or two male pronuclei (Figure 5.3). Zygotes with two female pronuclei are termed **gynegenotes** and those with two male pronuclei are termed *androgenotes*. Such reconstructed mouse eggs have been used in several types of experiments designed to understand the respective contribution of the maternal and paternal genes to the development of offspring.

Both gynegenotes and androgenotes undergo development to blastocysts with nearly the same frequency as bisexual zygotes. The resulting blastocysts also implant in the lining of the uterus (discussed in Chapter 8) and undergo early organ development at nearly normal rates, but neither type of unisexual embryo gives rise to live offspring. Gynegenotes exhibit relatively normal development of the embryo proper but retarded development of extraembryonic membranes, including the placenta. Androgenotes exhibit retarded development of the embryo but relatively normal development of extraembryonic membranes and placenta. Thus, unlike the oxen, the work of developing an offspring requires not just a cooperative pair, but one male and one female member to the team.

There are two possibilities for androgenotes because the zygote can contain either an X or a Y chromosome from the father, depending on which type of sperm fertilized the egg. If it contains a Y chromosome, the offspring is male; an X chromosome leads to a female. This suggests that differences in developmental potential could be influenced by the chromosome makeup of the

Gynegenote: a cleaving egg with two female pronuclei at the zygote stage.

FIGURE 5.3
Exchange of pronuclei between zygotes.

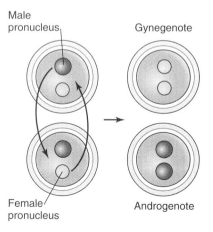

Male pronucleus

Gynegenote

Female pronucleus

Androgenote

androgenote, that is, XX, XY, or YY. In fact, the chromosome composition of the androgenotes had little influence on embryonic development; what was clearly important was the absence of the female pronucleus. This was a puzzling observation because the chromosome composition of a heterosexual zygote could also be either XX or XY, with the X chromosome always being derived from the mother and the other X or Y chromosome being derived from the father. A homosexual gynegenote would always be XX. The difference in developmental potential of the zygotes thus appeared to result from the origin of the X chromosome: ovary or testis.

Several types of studies have since confirmed that the expression of a gene is not only determined by its position on the chromosome but may also be influenced by whether it passed through an ovary or a testis. This type of genetic influence is termed *imprinting*. Although all the mechanisms of genetic imprinting are not known with certainty, one possibility is the addition of a methyl (–CH3) group to one of the DNA molecules (cytosine, described in Chapter 7). This process is termed *DNA methylation* and does not alter the fundamental sequence of the gene but appears to influence whether or not the gene is expressed and under what circumstances. Such modification of a gene is termed **epigenetic** to indicate that the gene sequence itself is not altered. Although there are some exceptions, as a rule regions of chromosomes whose genes are methylated on many cytosine residues (**hypermethylated**) are not expressed. Removing the methyl groups may allow the gene to become expressed. Several fundamentally important imprinted genes have been identified in the mouse.

Epigenetic: a modification to a gene that does not involve changes in the sequence of the bases.

Other types of experiments have attempted to understand communication between the pronuclei and the cytoplasm. A particularly interesting experimental system is a mouse strain, DDK, that exhibits mating specificity. Females mated to mouse strains other than DDK males exhibit very low pregnancy rates, although their eggs undergo fertilization and one or two cleavage divisions. Other female strains mated to DDK males exhibit normal fertility. Series of reciprocal pronuclei crosses (Figure 5.3) between strains has revealed a specific interaction between DDK male pronuclei and DDK egg cytoplasm that is essential for fetal development. Transfer of DDK egg RNAs into zygotes of other strains results in the same type of embryonic arrest at the morula/blastocyst transition observed for DDK eggs fertilized with "foreign" sperm. This result suggests the existence of maternal messages in the DDK egg that are compatible only with DDK sperm. Mating studies have indicated the gene involved (termed *ovum mutant*) is on chromosome 11 of inbred laboratory strains of mice, probably within a gene cluster termed *Schlafen*. Understanding the nature of this interaction may provide important clues about the continuing role of egg cytoplasmic stores in gene expression during early development. This is an example of a specific genetic trait in a laboratory mouse providing clues about a fundamentally important developmental process. This may be especially important to lines of communication between egg cytoplasm and nuclei in general, including nuclei transplanted for the purpose of making stem cells.

In summary, the zygote is the result of the remarkable ability of the egg to reorganize chromosomes into pronuclei. It seems likely that factors that can remodel both sperm heads and metaphase II egg chromosomes also can be used to remodel the nuclei from somatic cells, as discussed in Chapters 7 and 10. With respect to reproduction, parental DNAs are maintained as separate sets of chromosomes throughout the duration of the zygotic cell cycle, until the first metaphase occurs. It is the two-cell stage, not the zygote, that commingles parental chromosomes for the first time to define a new individual. Thus, "life begins at the two-cell stage" is more accurate than "life begins at fertilization." However, the paternal genes may not be expressed and thus play no role in development at the two-cell stage. A better definition of "new life" may be when the paternal genes play a role in guiding development. This is not known with certainty for most species. The best available evidence suggests the paternal genes are expressed for the first time at the four- to eight-cell stage of human development, as discussed in Chapter 6.

Parthenotes

As discussed in Chapters 1 and 4, parthenogenetic activation of human eggs could be of particular value in the derivation of human embryonic stem cells. Two types of parthenotes have been studied, principally from mouse and rabbit eggs. One type is diploid and the other haploid (see Figure 4.2). Egg activation is brought about by a variety of external stimuli, including electrical impulses and chemicals, termed *ionophores*, that stimulate release of calcium stores.

Haploid Parthenotes

A haploid parthenote results from egg activation with extrusion of the second polar body. Strictly speaking, this type of parthenote would be a haploid gynegenote. Studies of haploid mouse parthenotes revealed that early cleavage divisions can occur with a haploid number of chromosomes, indicating that the zygotic cell cycle does not depend on the full diploid complement of chromosomes. As discussed previously, this may be because checkpoints for DNA replication are not as rigorous in the zygote as they are in somatic cells. This circumstance may be necessary to allow entry into G2 and M phase after replication of the haploid number of chromosomes in each pronucleus.

Importantly, stem cell lines have been derived only from haploid parthenotes that have become diploid during subsequent cell divisions. For this to occur, a cycle of DNA replication with no cell division had to take place, presumably within the first few cleavages. Thus, the haploid nature of the genome may be sensed in subsequent cell cycles. If diploid parthenote stem cell lines could be derived from activated human eggs, it could provide a source of well-defined stem cells of a single set of genetics without the social stigma and more complex genetics associated with stem cells derived from human embryos.

Another type of haploid parthenote has been studied to determine the role of egg chromosomes in remodeling the sperm head. Activated mouse eggs with the egg metaphase plate removed (termed *enucleated*, see Chapter 7) have been

Transgenic Mice
Introduction of Foreign Genes into Zygotes

The advent of remote-controlled devices attached to microscope stages that could manipulate tools within a range of a few microns paved the way for new types of experiments in several fields of biology. Originally developed to inject compounds that could not pass through cell membranes directly into the cells, the devices were quickly adapted to embryo manipulations by developmental biologists.

Two groups of scientists applied this technology to mouse zygotes in the 1980s with the goal of understanding controls on gene expression in mice. Holding the zygote by mild suction with a polished glass tube approximately 30 microns in diameter (the holding pipette), the pronucleus of the mouse zygote presents a clearly visible target for another glass tube, pointed, sharpened, and less than 10 microns in diameter (the injection pipette). This micromanipulation system provides the opportunity to introduce substances, including DNA sequences, directly into the pronuclei of zygotes. The obvious question was what would happen? Did the mouse embryos have the same "foreign DNA" sensing mechanisms as bacteria? If so, the injected sequences would be degraded. Amazingly, not only were the sequences not instantly degraded, but in many cases they persisted into the offspring. This acceptance of new DNA and plasticity of the mammalian genome was unexpected.

In 1981, Ralph Brinster, working with Richard Palmiter, microinjected a DNA construct containing the gene for herpes virus thymidine kinase into one pronucleus of mouse zygotes, transferred them back into oviducts of female mice for development, and demonstrated that not only had the DNA construct been carried forward into the developing mouse, but that some tissues expressed the thymidine kinase enzyme activity. This team followed this experiment in 1982 with a dramatic demonstration of the effects of injecting the gene for growth hormone into a mouse zygote. Of the 21 offspring that survived the procedure, 7, approximately 33%, carried the injected "transgene." That percentage of "transgenic mice" produced from microinjection experiments has remained consistent in subsequent experiments in many other laboratories. Of the seven animals that were transgenic for human growth hormone, six grew dramatically larger than their littermates.

In 1982, John Gordon and Frank Ruddle also developed the techniques for DNA microinjection into mouse pronuclei. These investigators were the first to demonstrate that some of the transgenic animals could transmit the new genes to progeny, in what appeared to be standard Mendelian genetics. Animals with the new genes incorporated into either their oogonia or spermatogonia were termed founders.

That new genes could be so easily inserted into the germ line not only has far-reaching consequences for our views of evolution and inheritance but provides an unprecedented and powerful opportunity to study controls on gene expression, as well as the influence of specific genes on cell functions. Thousands of transgenic mice have been created in laboratories throughout the world in the 20 years since the technology became available.

In 1989, an Italian team of investigators, headed up by Corrado Spadafora, demonstrated another approach to creating transgenic mice. They reported that exposure of sperm to solutions of DNA constructs allowed the incorporation of the DNA into the sperm head and then into the egg at fertilization. This report could not immediately be confirmed by other scientists, including Ralph Brinster and Davor Solter, and doubt was cast on the reliability of the information reported by Spadafora's research team. However, several other teams of scientists throughout the world have now reported sperm-mediated DNA transfer in several species, including fish, amphibia, birds, and a few mammals. Creating transgenic mice by this approach remains more difficult than with other species for unknown reasons and is an active area of research.

Given the potential exposure of sperm to foreign DNAs in transit to fertilization of an egg, it is alarming to consider the possibility that those foreign DNAs could be incorporated into the human genome. This could hold particular relevance to programs of assisted reproduction throughout the world in which fertilization occurs under a wide variety of laboratory conditions.

injected with sperm. Remodeling the sperm head into a male pronucleus proceeded at a normal rate in the enucleated mouse eggs, confirming the notion that egg cytoplasm contains stockpiles of cell cycle factors needed for nuclear remodeling and DNA replication. Strictly speaking, injecting a sperm into an enucleated egg is not fertilization, because the egg genetic information is gone. Presumably, stem cell lines would be difficult to derive from such haploid androgenotes, because sperm contain either X or Y chromosomes but not both. Thus, individual sperm do not represent the genetic composition of the man producing them.

Diploid Parthenotes

A diploid gynegenote parthenote arises from artificial egg activation before the second polar body is extruded, thus maintaining all 46 chromosomes of the metaphase II egg (see Figure 4.2). Under the right conditions, diploid gynegenotes can give rise to stem cells, as reported by Jose Cibelli and his colleagues for eggs from the monkey, *Macaca fascicularis*. Primate parthenotes undergo developmental arrest in utero, as described previously for mouse zygotes, and are therefore not really embryos. Pluripotent stem cells derived from diploid activated human eggs may therefore also be free from the social stigma associated with deriving stem cells from human eggs fertilized with sperm.

Diploid gynegenotes develop more reliably than do haploid parthenotes. Suppression of the extrusion of the second polar body by the activated egg is usually accomplished by culturing the egg in a common inhibitor of microtubule formation, as described below and in Chapter 4. In theory, it may also be possible to create a diploid androgenote from one X sperm and one Y sperm. Because this type of parthenote presumably would also not be capable of development, analogous to the mouse androgenotes, such androgenote parthenotes might also be more socially acceptable because they are not truly embryos. Work in this area is limited. A series of elegant experiments by Tomohiro Kono and his colleagues, described in Chapter 10, used the developmental failure of parthenotes to help define the contribution of sperm to embryogenesis. This information is important not only to understanding gene expression important during early development, but also how to efficiently derive pluripotent stem cells from parthenotes.

Once an interesting experimental tool for developmental biologists, parthenotes may become an especially valuable tool for stem cell medicine. Given the many ways that eggs may be activated without the potential to give rise to offspring, it becomes clear that new terms are needed to describe the processes. One principal difference between activating eggs to derive embryonic stem cells and activating eggs to give rise to an embryo may be the absence of the creation of new germ cells with new genetics. Thus, the eggs are being activated to give rise only to new somatic cells to treat the diseases outlined in Chapter 12. A new term, **ovasome**, has been proposed to replace the term *embryo* when the goal of activating the eggs ("ova") is to create new somatic cells ("somes"). The process, whether by parthenogenesis or by nuclear transplantation, to be described in Chapter 7, could be termed *ovasomagenesis*.

Ovasome: a new term to describe egg activation for the purpose of creating stem cells rather than an embryo.

Molecular Biology of Zygotes

Once the pronuclei have formed, the molecular pathways that controlled the arrest of the egg either at the germinal vesicle stage or in metaphase II have been eliminated. In addition, the cytoplasmic factors capable of remodeling the sperm head and forming nuclei have also been eliminated. Several investigators have discovered that either sperm or somatic cell nuclei, injected into zygote-stage eggs, do not undergo the nuclear remodeling characteristic of eggs after 2 hours of activation. Thus those components in the activated egg responsible for nuclear remodeling are degraded within a few hours. It will be important to nuclear transplant stem cell derivation to fully understand egg factors responsible for nuclear remodeling, as discussed in Chapter 7.

The giant cell enters the G1 to S transition, with DNA synthesis initiated independently in both pronuclei. Although a few other cells (e.g., placenta trophoblast, discussed in Chapter 8) contain multiple nuclei, they are formed by nuclear division without cell division. The need to coordinate DNA replication in two haploid nuclei simultaneously is peculiar to the zygote.

Because, in general, progression through the zygote stage is blocked in the presence of inhibitors of protein synthesis but not inhibitors of messenger RNA (mRNA) synthesis, it is assumed that protein synthesis, perhaps the cyclins and some members of the DNA replication complex, are required and that their synthesis is guided by mRNA stockpiles within the egg. There is some evidence that paternal genes may be transcribed during the zygote stage, but they have not been thought to be essential for cleavage to the two-cell stage. This concept has recently been challenged, however, by the discovery that gene transcription occurs at the beginning of S phase in a common laboratory mouse embryo and is essential to cleavage beyond the two- to four-cell stage. This is the earliest reported gene to be expressed after fertilization. It is a member of the repetitive elements now known to be ubiquitous in mice, some of which have characteristics of retrovirus elements. The zygotic gene expressed is a murine endogenous retrovirus, family L, MERV-L. Its role in early development is unknown. It might be expressed by the paternal pronucleus, which has been shown to have fewer constraints on gene expression than the maternal pronucleus in the zygote.

Changes on controls of gene expression in the zygote are not dependent on the completion of S phase or cell cycle progression, however, but are more related to the time from egg activation. A type of clock is started, perhaps by the calcium ion pulses described in Chapter 4. Some cell functions, such as changes in the controls on gene expression, occur even if S phase is incomplete or cell division is blocked. Important aspects of egg activation are independent of the nature of the genes within the egg cytoplasm. This is of obvious importance to the ability of the egg to reprogram gene expression from the chromosomes of adult somatic cells, as discussed in Chapter 7.

As discussed previously, the G2 phase in the zygote is short. Once the pronuclei meet in the center of the egg, approximately 20 hours after sperm entry, the G2 to M transition commences. An area of fundamental importance to the first cleavage is organization of the centrosome, the cytoplasmic organizer that defines the cleavage poles of the egg. In the mouse, the egg contains two centrosomes, but in the human it is the sperm that brings the centrosome into the egg. The centrosome (Figure 5.4) is a structural protein assembly in the cytoplasm. It is a cylindrical protein core, the centriole, surrounded by a matrix of proteins, such as gamma-tubulin, that assemble into the microtubule motor responsible for anchoring the chromosomes and separating them into two new daughter cells. Spindles are the structural proteins that attach to the chromosomes and serve to precisely drag one of each into the new daughter cell. The centrosome assembly at the periphery of the separating chromosomes resembles a star and is therefore referred to as **astral spindle**.

Many of the details of centrosome biology are unknown. For example, it is known that the centrosome undergoes replication in the cytoplasm during S phase when DNA replication duplicates the chromosomes in the nucleus. This provides a centrosome for each end of telophase. The centrosome appears to send out the structural proteins that anchor to the chromosome region, the **kinetochore**, used to pull them apart.

There are, however, also examples of cell division in the absence of centrosomes, termed *anastral spindles*. Such spindles form in an "inside out" fashion from the kinetochore of the chromosome out to the structural protein anchor that serves to pull the chromosomes apart. This mechanism may be less common than astral spindle assembly, but research is lacking.

These considerations are especially important to the biology of human eggs that do not appear to contain a centrosome with a centriole. Human parthenotes

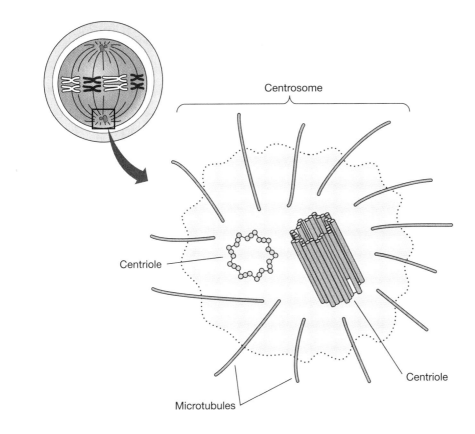

FIGURE 5.4
The centrosome.

Centrosome

Centriole

Centriole

Microtubules

need to divide without a sperm centrosome, and information is lacking about the transition of cells from anastral to astral spindle formation. Would human parthenote stem cells always divide without a centriole? Or would centrioles ultimately be produced after activation and cleavage?

This is also an important consideration for nuclear transplantation in human eggs: how to bring about successful cell cleavage in the absence of sperm centrosomes. Is centrosome formation an important part of nuclear reprogramming? This is an area in need of additional investigation, as discussed in Chapter 10.

A peculiar characteristic of zygotes, including mouse and human, is their dependence on exogenous pyruvate as an energy source for cleavage. This fundamental observation was made by Roy Whitten in 1957 as he attempted to define laboratory culture conditions that would support early development of mouse and rabbit embryos. He had observed development to blastocysts in culture by two-cell mouse embryos, but zygotes were blocked before cleavage. The inclusion of pyruvate in the culture medium solved the problem. More recent work has shown that pyruvate is necessary to support protein synthesis in the zygote, supporting the concept that protein synthesis is necessary for progression through the first cell cycle. Many of the metabolic pathways of zygotes and blastomeres have not been described, but the requirement for pyruvate, and the lack of utilization of glucose for energy, may relate to the quiescent nature of oocyte mitochondria.

As discussed above, the transition from germ cells to embryonic cells requires changes in gene expression. The advent of high-throughput gene sequencing methods has allowed the compilation of libraries of mRNAs expressed at each stage of development. Taken together, mRNAs from 9,718 genes were detected for unfertilized eggs, zygotes, 2-cell, 4-cell, 8-cell, 16-cell, and blastocysts. Surprisingly, half of the genes were novel, revealing that a considerable fraction of mouse genes are dedicated to embryonic expression. Interestingly, 243 genes were identified as being specifically expressed in mouse zygotes, suggesting they may play key roles in the transition from germ cells to undifferentiated blastomeres. Not all the gene products have been identified, but it is important to become familiar with some of the more abundant ones, as listed in Table 5.1. The 243 genes were assigned to zygotes

TABLE 5.1	Genes Expressed Specifically in Zygotes
GenBank Number	**Gene Product**
J04596	PDGF-derived growth factor-inducible KC protein
M12253	Alpha-tubulin isotype M-alpha-6 mRNA
M64085	spi2 proteinase inhibitor
S43105	Cyclin B1
M84361	Macrophage colony-stimulating factor (CSF-1)
X55957	Inhibin alpha subunit
U87557	Phosphatidylcholine-specific phospholipase D2
X71327	MRE-binding transcription factor
X78682	B-cell receptor associated protein (BAP) 32
U01063	pLK serine/threonine kinase
U13838	Vacuolar adenosine triphosphatase subunit B
U38690	DAZ-like putative RNA binding protein
D16333	Co-proporphyrinogen oxidase
U36757	Thrombin receptor (Cf2r)
AF001688	U4/U6 snRNP 90 kDA protein
D38517	Dhm1 protein
M29324	LINE-1, Md-A13
U48972	Spindlin
U04672	Bone morphogenetic protein receptor kinase-I

because they were detected at markedly higher frequency among the mRNAs isolated from zygotes than from other developmental stages, described in Chapter 6.

It is important to discover which genes are important to pluripotency and which to future differentiation into functioning organs. The pluripotency genes may play critically important roles in stem cell behavior and thus may need to be expressed anew in somatic cells transplanted into eggs for the derivation of patient-specific stem cells, as discussed again in Chapters 7 and 10.

Summary

The zygote is a unique stage of development that occurs once for each egg and lasts approximately a day for human eggs. The giant cell is stockpiled with the molecular machinery to reprogram gene expression by sperm, egg, and somatic cell chromosomes, described in Chapter 7. Not surprising, successfully reprogramming sperm and egg chromosomes is far more efficient than successfully reprogramming somatic cell chromosomes. The more we understand about an activated egg's first day, the greater the possibility of harnessing this remarkable natural power.

A fundamentally important task for the zygote-stage egg is faithful replication of the chromosomes within its cytoplasm, be they from sperm, egg, or somatic cell. The complex cellular machinery needed for this task is stockpiled within the egg and may be deployed by the periodic calcium releases that accompany egg activation. Details about these processes are urgently needed.

Organization for the first cleavage division to two cells requires communication between egg cytoplasmic factors and the newly activated nucleus, but the exact nature of that communication is not known with certainty. Advances in the technology to detect mRNAs in a small number of cells have provided valuable new clues about genes that are expressed specifically in the egg and in the zygote.

Review Questions (True or False)

1. The zygote contains an unusually large nucleus, termed a *pronucleus*.
2. The chromosomes within the pronucleus have undergone remodeling to a state of gene expression more characteristic of pluripotent stem cells than of egg chromosomes.
3. Faithful replication of the DNA in the chromosomes of the pronucleus is an important zygotic activity.
4. DNA replication involves a large complex of proteins that come together in random areas of the chromosomes.
5. A gynegenote contains two sperm heads that have been remodeled into pronuclei.
6. Imprinting is an example of an epigenetic modification of DNA.
7. DNA injected into the pronucleus is immediately degraded by pronuclear enzymes.
8. All genes expressed in the zygote are also expressed in the egg before activation.

ADDITIONAL READINGS

Battaglia, D. E., Klein, N. A., and Soules, M. R. (1996). Changes in centrosomal domains during meiotic maturation in the human oocyte. *Mol Hum Reprod* 2: 845–851.

Gao, S., Wu, G., Han, Z., Casa-Esperon, E., Sapienza, C., and Latham, K. (2005). Recapitulation of the ovum mutant (Om) phenotype and loss of Om locus polarity in cloned mouse embryos. *Biol Reprod* 72: 487–491.

Kiessling, A. A. (2005). Eggs alone. *Nature* 434: 145.

Ko, M. S. H., Kitchen, J. R., Wang, X., Threat, T. A., Wang, X., Hasegawa, A., Sun, T., Grahovac, M. J., Kargul, G. J., Lim, M. K., Cui, Y. S., Sano, Y., Tanaka, T., Liang, Y., Mason, S., Paonessa, P. D., Sauls, A. D., DePalma, G. E., Sharara, R., Rowe, L. B., Eppig, J., Morrell, C., and Doi, H. (2000) Large-scale cDNA analysis reveals phased gene expression patterns during preimplantation mouse development. *Development* 127: 1737–1749.

Monk, M., ed. (1987). *Mammalian Development: A Practical Approach*. Oxford: IRL Press.

Renard, J. P., Baldacci, P., Richoux-Duranthon, V., Pournin, S., and Babinet, C. (1994). A maternal factor affecting mouse blastocyst formation. *Development* 120: 797–802.

Turner, B. M. (2000). Histone acetylation and an epigenetic code. *Bioessays* 22: 836–845.

Varmark, H. (2004). Functional role of centrosomes in spindle assembly and organization. *J Cell Biochem* 91: 904–914.

Wassarman, P. M., ed. (1991). *Elements of Mammalian Fertilization*, Vols. I and II. Boca Raton, FL: CRC Press.

Wiekowski, M., Miranda, M., and DePamphilis, M. (1993). Requirements for promoter activity in mouse oocytes and embryos distinguish paternal pronuclei from maternal and zygotic nuclei. *Dev Biol* 159: 366–378.

CHAPTER 6

Blastomere Cleavage

It is this cell which plans and composes all organisms, and which transmits to them its defects and potentialities.

Albert Claude

OVERVIEW

Once activated, the egg initiates a series of unique cell cycles, each of which results in identical daughter cells with smaller volumes. The reduction in size is the reason these cell cycles are termed *cleavages* rather than cell divisions. Before programs of human in vitro fertilization (IVF) became standard of care for infertile couples, the term *embryo* was not applied to early cleavage stages. These stages were referred to by developmental biologists as cleaving eggs (see sidebar, Chapter 4). The term *cleaving eggs* was confusing to infertility patients, however, because it did not adequately describe whether or not the egg had been successfully fertilized before it started to cleave. Thus, the term *embryo* became commonly used to refer to cleaving eggs that had exhibited the zygote stage after exposure to sperm.

The use of the term *embryo* for these early egg cleavage stages has led to substantial confusion, however, about the developmental potential of the fertilized egg. Moreover, the term *embryo* suggests that the newly compiled complement of genes is controlling early cleavage events. This is not the case. Several lines of evidence make it clear that the egg is responsible for guiding early cleavages. For example, they take place with or without the process of fertilization, as described in Chapters 4 and 5. Exactly when the new complement of genes takes over control of each cell is not known with certainty. It appears most likely to be a somewhat gradual process that occurs over several cell cycles. For accuracy, the term *embryo* should only be applied to those developmental stages in which the newly compiled embryonic genes are controlling cell functions. Those individuals wishing to define precisely when a new life begins should perhaps reserve that definition for the "embryonic **genome** control" stage, which is not known with certainty for human eggs at this time. Additional research is needed.

Genome: a general term for all the genetic information of an organism.

The First Cleavage

The first cleavage occurs approximately 22 and 26 hours after exposure of mouse and human eggs to sperm, respectively. Parthenogenic stimulation leads to approximately the same time course.

However activated, the cleaved human egg gives rise to two smaller cells approximately 80 microns in diameter, each comprising approximately 270,000 cubic microns. Careful measurements of mouse embryos have revealed relatively little change in mass with respect to overall weight, protein, lipid, and nucleic acid content throughout early cleavages. Assuming the same for human eggs, the contents of 800,000 cubic microns in the human egg (see Chapter 3) are therefore divided between two cells with a combined volume less than that, resulting in an increase in concentration of cytoplasmic contents. In contrast, the surface area of the egg, approximately 42,000 square microns, is divided approximately evenly between the two blastomeres, each with a surface area of approximately 20,000 square microns. The net effect of the first cleavage is therefore an increase in the ratio of plasma membrane to cytoplasm (Table 6.1) whose components, including signal transduction pathways, have undergone concentration. Whether or not this increased ratio of cell surface to cytoplasm influences the blastomere's response to its environment through cell surface sensors is a matter of speculation.

As mentioned in Chapter 4, a unique physiologic requirement of the zygote to undergo the first cleavage is the presence of pyruvate instead of glucose for energy. This fact was discovered by Roy Whitten, a New Zealand scientist, while trying to understand why two-cell stage eggs recovered from mice for culture in the laboratory would develop for 4 more days, to the blastocyst stage, whereas one-cell stage eggs frequently arrested in the same culture conditions. In an effort to measure culture medium components with as much precision as possible, Whitten turned to calcium pyruvate as a calcium salt, which did not spontaneously attract as much water during storage as did other calcium salts, such as calcium chloride. Amazingly, in this new medium, zygotes cleaved to two cells and then on to four cells, thus establishing that unlike most somatic cells, pyruvate is essential for the first cleavage division. In most cells, glucose is metabolized to pyruvate in the cell cytoplasm, and the resulting pyruvate is used to generate

TABLE 6.1	Physical Dimensions of Early Cleavage Stages			
Stage	Diameter (microns)	Volume (cubic microns)	Surface Area (square microns)	Ratio of Surface Area to Volume
Egg	115	800,000	42,000	0.05
Two cell	80	270,000	20,000	0.07
Four cell	60	112,000	11,000	0.1
Eight cell	42	39,000	5,500	0.14
Sixteen cell	30	14,000	2,750	0.2
Thirty-two cell	21	5,000	1,400	0.28

energy (Figure 6.1). The zygote and early cleavage blastomeres may therefore not contain the enzyme pathways for the initial steps of glucose metabolism. Because one of the first steps in glucose metabolism requires more energy than it creates, perhaps there is a developmental advantage to not using precious energy stores in the egg for glucose metabolism. It is also possible that highly specific energy requirements may play a role in the extraordinarily long developmental arrest of mammalian eggs. Cell cycle energy considerations are fundamentally important to the design of laboratory conditions to support egg activation and early cleavages, as is discussed later in this chapter.

Understanding the sequence in which genes are deployed during early cleavages may reveal key steps in reprogramming eggs (and sperm) into the totipotent cells of the early embryo. The libraries of cloned messenger RNAs (mRNAs) created for each early cleavage stage of mouse embryos have revealed stage-specific expression of some genes, as described in Chapter 5 for zygotes. Interestingly, some genes were found to be expressed as mRNA for two or three cycles, but not more. Examples of genes expressed in the zygote and the two-cell stage are listed in Table 6.2.

It is important to discover why these particular genes are turned on in the zygote, persist in the two-cell stage, but are not detected as mRNAs at later stages in the preimplantation embryo. For example, CDC42 is known to be involved in assembly of proteins that define the cell skeleton and is important to early frog embryo development. It could play an important role in stabilizing the huge blastomeres of the two-cell stage and either not be needed during later cleavages, or the protein could have such a long life in blastomeres that continued expression of mRNA is not necessary. This is an important consideration for stem cell derivation. As described in Chapters 8 and 10, pluripotent stem cells capable of continual self-renewal have not been isolated from early cleaving embryos, only from stages past the 32-cell stage when the cells are smaller, the same size

Energy consumed

Glucose

CH₂O (P)
C
H
C   OH   H   C
HO   C   C   OH
H   OH

Glucose-6-phosphate

CH₂O (P)
C   O   CH₂OH
C   H   HO   C
H   C   C   OH
OH   H

Fructose-6-phosphate

CH₂O (P)
C   O   CH₂O (P)
C   H   HO   C
H   C   C   OH
OH   H

Fructose-1,6-bisphosphate

CH₂O (P)
C=O
CH₂OH

Dihydroxyacetone phosphate

⇌

CH₂O (P)
H—C—OH
C=O
H

Glyceraldehyde-3-phosphate

Energy produced

FIGURE 6.1
Glucose metabolism.

FIGURE 6.1
(Continued).

Aerobic conditions

Anaerobic conditions

TABLE 6.2	Genes Expressed Specifically in Mouse Zygotes and Two-Cell Stage Embryos

GenBank Number	Gene Product
U37720	GTP-binding protein, CDC42
U16818	UDP glucuronosyltransferase (UGT1-06)
U51907	TRAF family associated NF-kappa B activator (TANK)
X57413	Transforming growth factor-beta 2
S59342	Nuclear pore complex glycoprotein p62
U89506	Lark RNA-binding protein
U47737	Thymic shared antigen (TSA)-1
S82156	GST 5-glutathione S-transferase-sperm antigen MSAg-5 fusion protein
M35797	T-complex protein (Tcp-1x)

as somatic cells. Could this be due to failure to adequately support the geometry of the large earlier-stage blastomeres? Could over-expression of CDC42 stabilize the huge cells in laboratory culture? Additional research is needed.

Similarly, Lark is a protein important to normal development of fruit fly embryos that is involved in the rhythmic expression of some genes, reminiscent of the patterns of calcium ion release noted in the activated egg (see Chapter 4). It is therefore not surprising that it is expressed in early mouse embryos, but why only to the two-cell stage?

The Two-Cell Stage

Several lines of evidence support the developmental equivalence of each blastomere of the two-cell stage. For example, separating them and transferring each blastomere individually to the uteri of different mothers lead to identical twins in several mammalian species, including monkeys. Hence, each blastomere is capable of giving rise to both placenta and embryo. This is in stark contrast to frogs and insects, whose eggs are compartmentalized with respect to giving rise to head, tail, back, or belly from the very beginning of development.

Evidence that developmental equivalence of the first two blastomeres of mouse embryos may be an over-simplification has, however, grown in recent years. The controversy is ongoing. Magdalena Zernicka-Goetz at the University of Cambridge, Richard Gardner at the University of Oxford, and Michael Roberts at the University of Missouri have provided evidence that the mouse zygote has an animal–vegetal axis and the first cleavage leads to blastomeres with predetermined embryonic fates. Davor Solter of the Max Planck Institute disagrees and has provided evidence that the mouse egg does not have an animal–vegetal pole.

It seems possible that both points of view have merit and may relate to the concept of equivalence groups, discussed in Chapter 8. Briefly, the notion is that as cells enter commitment pathways (e.g., to becoming heart muscle cells), some plasticity remains during the early stages. For example, to ensure that a heart develops, perhaps twice as many cells enter that pathway as are actually needed. Successful development generates a feedback signal that the remaining cells in the "may-become-heart" group can now enter another pathway.

Similarly, given the size of mammalian eggs, cytoplasmic components may not be entirely uniform throughout. This could lead to some inequality in parsing developmentally important components to each of the two-cell blastomeres. Such inequality could, in turn, lead to a preference for a later position in the embryo. An example of such inequality is the transcription factor discovered in the laboratory of Michael Roberts to be concentrated in the two-cell blastomere that contributes most of the cells in blastocyst trophectoderm. Discovering the nature of the guiding components will provide valuable new insights into controls on early mammalian commitment events.

Importantly, however, inherent plasticity in each two-cell blastomere's developmental program ensures totipotency if needed. This is also an important concept for totipotent or pluripotent stem cells. Broad plasticity in their developmental potential is essential to their ability to continue to self-renew and remain capable of becoming committed to all developmental pathways.

The Second Cleavage

The two-cell stage marks the first commingling of maternal and paternal genes. In theory, the chromosome separation at the first cleavage event resulted in one maternal and one paternal member of each chromosome pair being allocated to each blastomere. As discussed previously, given the semiconservative replication of DNA, each new chromosome created in the zygote has one germ-cell strand of DNA and one zygote strand of DNA (see Figure 2.6). The G1 phase of the two-cell blastomere is short for both mice and human eggs, and S phase begins within a couple of hours of cell cleavage. Interestingly, but for reasons that are not clear, the mouse two-cell blastomere maintains a cytoplasmic bridge between the two blastomeres until shortly after S phase begins. Information is lacking about whether or not this situation is true for human two-cell blastomeres. The short G1 phase provides support for the concept that precursors for the DNA replication machinery preexist in the two-cell blastomere. The checkpoint for "start" may not exist or may be different from somatic cells. S phase requires 6 to 8 hours for completion, in keeping with the S phase of somatic cells.

Half of the chromosomes resulting from S phase in the two-cell blastomere are composed of double-stranded DNA in which both strands have been replicated after egg activation (see Figure 2.9). The other half of the chromosomes contain one new DNA strand and one "original" DNA strand contributed by the egg and/or sperm. As discussed in Chapter 5, genes are imprinted in germ cells. Genes in sperm are imprinted differently from the same gene in the egg. The principle mechanism of imprinting is thought to be controlled by methylation of cytosine within the DNA strand, so-called *epigenetic* regulation. Patterns of methylation control gene expression. Precisely how methylation patterns are reestablished in the newly synthesized DNA strand is not known with certainty, although there is evidence that the enzymes responsible for adding the methyl group, "DNA methylases," follow the pattern present on the opposite strand. Such epigenetic controls on gene expression are an active area of research in both developing embryos and in pluripotent stem cells.

The G2 phase of the mouse two-cell stage is unusually long, about 12 hours. The reasons for this are unclear, although many processes are known to occur during this phase, as discussed in the Molecular Biology section of this chapter. The long G2 of the mouse two-cell stage is also a period during which developmental arrest may occur, termed the *two-cell block*. A number of factors contribute to the two-cell block, including the genetics of the egg and the presence of glucose and some purines. A full understanding of this unusual cell cycle arrest will provide important insights into controls on blastomere cleavage.

Of the 282 genes detected specifically in the mouse two-cell blastomere, only 35 were known in GenBank, examples of which are listed in Table 6.3. The role of some of these genes appears to be obvious, such as myelin gene expression factor-2, probably important to the increased demand for myelin for blastomere division. But the role of others, specifically in the two-cell stage, remains to be discovered and will provide important clues about controls during cell division of pluripotent cells. Spontaneous separation of the two cells is one mechanism of twinning, including approximately one-third of identical human twins. This observation speaks against the notion that "life begins at fertilization," because in the case of the identical twins resulting from separation of early blastomeres, their individual lives obviously do not begin until that separation occurs. This observation also emphasizes an important distinction between mammalian embryos and those of lower animals, namely that each early blastomere maintains totipotency to develop into all tissues necessary for gestation and development into an offspring, including placenta and extraembryonic membranes. This is in contrast to insects and amphibians whose eggs are polarized and where distinct regions of the cytoplasm are precommitted to becoming head, tail, back, and belly. In contrast, therefore, mammalian embryos must rely on other signaling mechanisms to bring about cell commitment events that lead to tissue and fetal

TABLE 6.3	Genes Expressed Specifically in the Mouse Two-Cell Embryo
GenBank Number	**Gene Product**
D01034	Transcription factor IID (Tbp)
S43105	Cyclin B1
U08440	Cytochrome *c* oxidase subunit VIaL
U28016	Parathion hydrolase (phosphotriesterase)-related protein
L16846	B-cell translocation gene-1
NM008229	Histone deacetylase 2
U13262	Myelin gene expression factor-2
U58105	Alpha-D-galactosidase A, Bruton's tyrosine kinase
U10871	MAP kinase, p38b
NM010810	Matrix metallopeptidase 7 (Mmp7)
M60474	Myristoylated alanine-rich C-kinase substrate
X91144	P-selectin glycoprotein ligand 1
U57692	N-terminal asparagine arnidohydrolase (Ntan 1)
L11651	Translation initiation factor 5 (eIFf-5)

FIGURE 6.2
Cell cycle of mouse two-cell stage
blastomere and human two-cell
stage blastomere.

Mouse	G_1	S	G_2	M

Human	G_1	S	G_2	M

development. Although this is true, more recent evidence suggests that factors within the egg itself actually continue to influence fetal development. Unraveling these mysterious pathways is an urgently needed area of research, especially for human embryos.

Unlike the mouse, the G2 phase of the human two-cell blastomere is not unusually long, and cleavage to four cells occurs within approximately 12 to 18 hours (Figure 6.2). There is, however, a wide variation of cell cycle times for the early cleavages in the human, which is thought to relate to developmental potential. Many reports from infertility programs have revealed that pregnancy rates are substantially higher if the more rapidly cleaving embryos are selected for transfer into the uterus of the woman undergoing IVF. This has led to the notion that the developmental potential of more slowly cleaving embryos is reduced relative to the faster cleaving embryos, but many other variables could account for the variation in pregnancy rate, such as asynchrony of slower cleaving embryos with the uterus so that development is out of phase for the uterine lining (discussed in Chapter 8). Until testable parameters are established to reliably determine developmental potential, this will remain an area of speculation.

Cleavage of the two-cell blastomere to four cells occurs relatively synchronously, usually within a few minutes of each other for mouse embryos. Human two-cell blastomeres are somewhat less synchronous, however, as evidenced by the relatively common appearance of three cells. Each four-cell human blastomere is approximately 60 microns in diameter, with a volume of approximately 112,000 cubic microns, slightly less than half that of each two-cell blastomere and approximately one-seventh the volume of the egg (Table 6.1). The surface area of each human four-cell blastomere is approximately 11,000 square microns, on the order of half that of the two-cell blastomere. Thus, the second cleavage also results in an increased concentration of cytoplasmic components, as well as an increase in the ratio of cell surface to cytoplasm. These considerations suggest a conservation and reuse of plasma membrane with each cleavage division, along with increasing concentrations of cytoplasmic constituents. This strategy could serve to limit the need for synthesis of new components for the daughter blastomeres. Compacting the constituents in the cytoplasm could compensate for losses that may occur with each cleavage. For example, nucleosides and their phosphorylated forms essential for DNA synthesis have optimal concentrations for enzyme activity. Decreasing the volume of the cytoplasm could serve to maintain optimal concentrations of nucleosides in the face of stores depleted by each round of DNA replication.

The uniqueness of the first two cleavage stages is further emphasized by gene expression in the two-cell blastomeres that persists into the four-cell stage but not into the eight-cell stage. Examples of such genes are listed in Table 6.4.

TABLE 6.4	Genes Expressed in Two- and Four-Cell Mouse Embryos
GenBank Number	**Gene Product**
AF003346	Ubiquitin-conjugating enzyme, Ubcm2
D17571	NADPH-cytochrome P450 oxidoreductase
D89063	Oligosaccharyltransferase
J03298	Uterine lactotransferrin
X52046	Pro-alpha-I collagen type III

Fragmented Cleavages

An unusual difference between mouse and human blastomeres is the incidence of fragmentation in human blastomeres. The term *fragmentation* is used to describe the appearance of small blebs of plasma membrane–enclosed cytoplasm that are associated with early cleavage stages. It is as if part of a blastomere forms its own cleavage plane independent of the chromosomal alignment (Figure 6.3).

FIGURE 6.3
Fragmented human blastomeres.

The appearance of such fragments is relatively common in cleaving human embryos and relatively rare in cleaving mouse embryos.

The relative differences in fragmentation between the two species may be related to an interesting difference between the sperm of the two species. As described in Chapter 2, cell division is oriented according to the positioning of the centrosome, a cytoplasmic structure that contains the centriole that is the organizing center for the formation of the mitotic spindle. Interestingly, mouse eggs contain their own centrosome, but human eggs do not. Unlike mouse sperm, which lack a centrosome, human sperm contain two. It is speculated that abnormalities in sperm centrosomes could account for some human fertility problems and aberrant cleavages of fertilized eggs.

Recent time-lapse videos of cleaving mouse eggs, however, suggest another mechanism may also be operational (Figure 6.4). A cleaving two-cell blastomere was observed to undergo fragmentation followed by reannealing of the fragments. This observation emphasizes the plasticity of the membrane of the cleaving blastomere. That the human sperm centrosome is not essential for early cleavage events is, however, supported by several lines of evidence that parthenogenesis of human eggs does occur, both spontaneously in the ovary, as described in Chapter 3, and in laboratory conditions, as recently reported (see Chapter 1). It is important to determine whether parthenogenically activated human eggs exhibit a greater incidence of fragmentation during cleavage than eggs fertilized by sperm.

Analogous to the mouse two-cell blastomere, it is the human four-cell blastomere that is prone to developmental arrest (Figure 6.5). There is no evidence that the G2 phase of the human four-cell blastomere is as unusually long as the mouse two-cell blastomere, but arrest at the four-cell stage is commonly reported by infertility clinics. This may be related to the timing of the onset of blastomere gene expression, as described in the Molecular Biology section of this chapter.

Although mouse four-cells are not prone to developmental arrest, some genes are specifically turned on in the four-cell stage and do not persist into the eight-cell stage, examples of which are listed in Table 6.5. Similar information is not currently available for human four-cell stage embryos, but the need for specific, new gene expression could explain the propensity for developmental arrest of human four-cell stage embryos. For example, it may be highly significant that origin recognition complex protein 2 (mORC2L, see Figure 2.14) is synthesized first in the four-cell embryo. Either the egg stores of this protein were sufficient for the first two rounds of DNA synthesis or this protein was not required in the zygote and two-cell blastomere, suggesting novel controls on DNA replication. This is an area in urgent need of additional research, not only to provide a better understanding of chromosomal replication during early embryonic and parthenote development, but also to understand controls on DNA replication in transplanted nuclei, described in Chapters 7 and 10.

FIGURE 6.4
Phases of blastomere reannealing recorded by time-lapse videomicroscopy. The time between frames is approximately 7 minutes.

| Mouse | G₁ | S | G₂ | M | |
| Human | G₁ | S | G₂ | | M |

FIGURE 6.5
Cell cycle of mouse four-cell stage blastomere and human four-cell stage blastomere.

TABLE	
6.5	**Genes Expressed in Mouse Four-Cell Embryos**

GenBank Number	Gene Product
S80082	Murine leukemia virus gag
M63245	Amino levulinate synthase (ALAS)-H
U27457	Origin recognition complex protein 2 (mORC2L)
D17614	Protein theta-subtype 14-3-3
AF030343	Peroxisomal/mitochondrial dienoyl-CoA isomerase (ECH1p)
M64301	Extracellular signal-related kinase (ERK3)

The Third Cleavage

Morula: a cleaving egg with more than four blastomeres that has not developed a blastocoel cavity; usually 8 to 32 cells.

The blastomeres that do not undergo cell cycle arrest cleave again within 18 to 24 hours. The third cleavage heralds the onset of the **morula** stage of development and is not as synchronous as the second cleavage. Assuming that the paradigm that plasma membrane will be completely recycled persists with each cleavage, the eight-cell blastomere will have a surface area of approximately 5,500 square microns, a diameter of approximately 42 microns, and a volume of approximately 39,000 cubic microns (Table 6.1). New gene expression occurs at the eight-cell stage, such as TAX responsive element-binding protein, the exact function of which is unknown, as are the identities of the other mouse genes expressed specifically in eight-cell blastomeres.

The Fourth and Fifth Cleavages

The 32-cell blastomere is approximately the size of a human somatic cell (Table 6.1) and as such is nearly 1/200th the size of the egg. Assuming no loss of egg cytoplasmic organelles, they have become concentrated more than fivefold. It is at this stage that each cell must begin to grow in size before division takes place.

It is interesting to speculate that the strategy behind the enormous size of a mammalian egg is to support approximately 16 replications of the chromosomes before each cell must assume responsibility for continuing cell division. Because each cell may remain totipotent, this paradigm allows for the loss of some blastomeres, perhaps because of errors in DNA synthesis, without jeopardizing the formation of a complete individual. Several lines of evidence have shown that removal of a few blastomeres at the 8- to 16-cell stage (embryo biopsy, Figure 6.6) does not compromise the developmental potential of the embryo. It thus seems reasonable to view the egg as an enormous self-contained DNA replication factory, with the corollary abilities to remodel the sperm head and muster cell cycle machinery.

FIGURE 6.6
Biopsy of four-cell and eight-cell embryos.

The notion that the egg is designed to be self-contained for several cycles of DNA replication and cell division suggests it does not need the exogenous growth signals associated with progression through G1 and into S phase, as described in Chapter 2. Moreover, the well-documented toxicity of serum to early stage embryos in culture, particularly the mouse, suggests that growth factors may, in fact, derail the well-orchestrated egg programs for progression through early blastomere stages. In this regard, it is important to note that anatomically these stages of embryonic development occur in separate compartments of the female reproductive tract. All well-studied mammalian species, for example, mouse, rabbit, cow, goat, sheep, hamster, and human, have oviducts (also termed *fallopian tubes*), which link the uterus to the ovary. The transit time through the oviduct coincides approximately with the stages of early cleavage divisions. Most embryos enter the uterus at the morula stage, as discussed further in Chapter 8. These characteristics hold important clues for the design of laboratory conditions to sustain these early cleavages, as discussed in the following section.

Embryo Development in Laboratory Conditions

The ability to grow cells in the laboratory (termed *in vitro*, meaning "in glass") was recognized over 100 years ago as essential to understanding the nature of cells and their metabolic requirements. Louis Pasteur proved that round cake-pan-shaped dishes with loose fitting lids that overlapped the sides of the dishes could be used to culture bacteria in warm incubators without the bacteria escaping from one dish into another. Called Petri dishes or Petri plates, they were modified to have optically perfect bottoms to allow the observation of cells placed into the dishes through "inverted" microscopes. Like bacteria, the cells also cannot escape from one dish to the other. The observation of living cells through the bottom of their culture dishes remains standard laboratory practice and has allowed the development of techniques for manipulating individual cells, as described in Chapter 7. The next problem was to understand what laboratory conditions would best support cell growth and division.

Maintaining a balance of acids to bases in blood is critical to sustaining life. This is because enzyme reactions require strict adherence to a narrow window of balanced acid/base conditions, measured as the concentration of hydrogen ions in blood. The measurement is performed with specially designed probes sensitive to hydrogen ions. The most commonly used term to describe the concentration of hydrogen ions is **pH**, a term related to the logarithm of the hydrogen ion concentration. Normal physiologic pH values for blood are 7.2 to 7.4. To maintain this narrow range of pH, blood and other body fluids contain molecules that can acquire or release hydrogen ions as needed to maintain the balance. Such molecules are termed *buffers*, an important example of which is the bicarbonate buffer system. Carbon dioxide is a gas that is a little bit soluble in water. If added to water under pressure, once the pressure is released, gas bubbles reform. This accounts for the bubbles in all *sodas*, a term derived from sodium bicarbonate. In solution, carbon dioxide can combine with water to form carbonic acid, which readily ionizes in the presence of sodium chloride (the most common salt in blood) to form sodium bicarbonate (Figure 6.7).

Carbon dioxide is a byproduct of many normal physiologic reactions, such as the metabolism of glucose (Figure 6.1). The cells in the lung are designed to release the carbon dioxide into the air when we exhale. Exactly how much is released and how much is retained is determined by the pH in the blood at the moment of exhalation. The monitoring system is highly dynamic. A normal concentration of bicarbonate in the blood is 23 to 25 millimolar (mM). To maintain a pH of 7.4 in blood, it is necessary to have a concentration of CO_2 equivalent to 6.5% of the air in the lungs. Because the concentration of CO_2 in the atmosphere is on the order of 0.2%, it is clear that inhaling does not provide much buffering capacity. How, then, can one maintain a normal physiologic pH for cells cultured in the laboratory? Once the need was recognized, chambers were designed with carbon dioxide flowing into them in measured amounts mixed with air. Culture fluids (termed *culture media*) were formulated with 23 to 25 mM bicarbonate to mimic as closely as possible the physiologic conditions provided by the blood.

Blood also contains a well-defined mixture of salts and other ions. The salt concentration in blood is termed *normal saline* and is responsible for maintaining the balance of water inside and outside cells. Cells placed in distilled water,

pH: denotes the negative of the log of the molar concentration of [H$^+$] in a solution; pH 7 indicates [H$^+$] concentration equal to 10^{-7} molar.

FIGURE 6.7
Bicarbonate buffering system.

$$H_2O + Na^{++}HCO_3^= \rightleftarrows CO_2\uparrow + H_2O + NaOH$$

for example, swell until they burst because water flows into the cell, attracted by the intracellular salt. It is not surprising, therefore, that the basis of all culture media is a salt solution very similar to that found in blood.

In addition, enormous efforts have been expended in defining and redefining the other ingredients in culture medium that are necessary to support many generations of identical cells. Starting with a single cell and allowing it to divide repeatedly is true "cloning" and can lead to the generation of millions of genetically identical daughter cells. This is clearly a more ideal experimental system than beginning with a heterogeneous population of cells from, for example, disaggregated tissues.

Initial studies to define the laboratory conditions that would support cell division revealed that a characteristic fundamental to cancerous tumors is that the cells grow readily in culture in a variety of conditions. More recent results have revealed that this is due to mutations in cell cycle checkpoint genes that allow cancer cells to escape normal cellular controls. For this reason, cancer cells in laboratory culture have escaped the routine cell cycle checkpoints for errors in DNA replication, as described for mutated p53 proteins in Chapter 2. Laboratory lines of cancer cells accumulate errors in DNA and even gain and lose chromosomes, a condition referred to as **heteroploidy.** Such cells are useful for some types of experiments, but not others. To study specific types of normal cells, scientists tried to understand the laboratory culture conditions necessary to support cell division of single normal cells. An important discovery in this regard was reported in 1948 by Dr. Katherine Sanford working in the laboratory of Dr. Wilton Earle at the National Cancer Institute (see sidebar). Dr. Sanford determined that single cells could continue dividing if they were cultured in medium "conditioned" by groups of cells, such as sections of whole tissue. In addition, by culturing single cells in small capillary tubes in a small amount of medium, she demonstrated that cells could condition their own medium if the volume was small enough. Not only did these experiments pave the way for expansion of normal cells important for many types of biomedical studies, but they suggested the existence of factors, produced by all cells, that stimulated cells to divide, so-called growth factors. The need for such growth factors was tested by culturing normal cell lines on top of feeder layers of cells that were living but could not divide anymore because their DNA had been extensively damaged by x-ray treatment. Such conditioned medium was analyzed by numerous laboratories to better understand growth factors, of which many families have now been identified. As described in Chapter 9, human embryonic stem cells have been derived by culture on feeder layers with carefully defined growth factor supplements.

The search for the best culture medium for early cleaving embryos is an active area of research. Considering the autonomous characteristics of the blastomeres, culture needs for early stages may be relatively simple, designed to avoid premature stimulation of growth factor pathways and maintenance of the internal stores of precursors for protein and nucleic acid synthesis. Pyruvate and/or lactate are the most striking single requirements known for early cleavages.

Heteroploidy: a noncharacteristic number of chromosomes, or parts of chromosomes, in a cell.

Development of Tissue Culture Media

Developing the formulations of buffers, salts, vitamins, and other nutrients that supported cell division in mammalian cells was the work of many investigators. Early work used relatively high concentrations of serum (30% of the volume), usually recovered from fetal calves at slaughter houses because it was naturally sterile, and the remaining 70% was supplemented with a wide variety of additives based on the composition of various body fluids. The systems produced wide variations in results and yielded little information about fundamentally important requirements for dividing cells. Two milestone articles published in 1955 changed the field radically. One was published by Dr. Theodore Puck and Philip Marcus at the University of Colorado in Denver, entitled "A Rapid Method for Viable Cell Titration and Clone Production with HeLa Cells in Tissue Culture." The goal was to culture HeLa, the cervical carcinoma cell line used in many studies, in a completely chemically defined medium like bacteria. The other study was published by Dr. Harry Eagle at the National Institutes of Health in Bethesda, Maryland, entitled "Nutritional Needs of Mammalian Cells in Tissue Culture." Together, these works laid the basis for understanding the important roles of amino acids, vitamins, salts (including trace metals such as iron), glucose, and abundant small peptides such as glutathione. Amino acids termed *essential* for survival were those the cells could not synthesize on their own; amino acids termed *nonessential* could also be synthesized by the cells. Those terms are still in use today as is the culture medium basal medium Eagle, BME.

Building on this basic medium formulation, numerous investigators have discovered specific requirements for specific cell types, thus yielding valuable information about the metabolic processes involved in various body tissues. Most hormones and growth factors have been discovered in this way.

Dr. Richard Ham joined the faculty at the University of Colorado to pursue his interest in culture requirements of cells growing individually instead of in large numbers. These studies have special relevance to cultured embryo cells, and it was Dr. Ham who formulated Ham's F-10 (he is now up to Ham's F-411) supplemented with pyruvate, which was necessary to promote cell division in Chinese hamster ovary cells, a cell line used in many experiments. A few years later, Dr. Whitten would discover that pyruvate was essential for the first cleavage in the mouse zygote. Thus it was not an accident that the first human IVF pregnancy, Louise Brown in England in 1980, resulted from fertilization of Mrs. Brown's egg in Ham's F-10.

Nonetheless, embryonic development in all available culture systems is not normal. This fact is well illustrated by mouse embryos. Because laboratory mice respond in a predictable fashion to exogenous administration of gonadotropins, development can be timed from the administration of the artificial luteinizing hormone surge (discussed in Chapter 3). Oocytes are released from ovarian follicles 12 hours after the injection of the hormone, and because the mice mate beforehand, the sperm are waiting in the oviduct. Thus, in vivo development of mouse embryos is just as synchronous as development after fertilization in the laboratory. For this reason, the mouse represents a useful mammalian system for testing the safety of embryonic culture conditions.

A comparison of mouse embryos developing in culture with those developing in pregnant mice reveals a marked lengthening of the second and third cleavages. This leads to in vitro embryos with one-half to one-third the normal number of cells at the time the first cell differentiation occurs, as discussed in Chapter 8. Whether or not somatic cell nuclei transferred into activated eggs require new formulations of culture medium for complete remodeling to a stem cell state awaits additional research.

Molecular Biology of Blastomere Cleavage

Sensitive techniques have detected mRNA synthesis in the G1 stage of the two-cell mouse embryo, but unlike somatic cells, the newly transcribed embryonic mRNA is not required for progression to S phase. In contrast, the mRNA synthesis that occurs during the long G2 phase of the two-cell blastomere is necessary for progression to four cells. The mRNA synthesis occurs even while oocyte mRNAs are being degraded. A peculiar feature of this cell cycle stage is profound degradation of residual oocyte mRNAs, as if in preparation for transition to embryonic genome control of cell functions.

The earliest transcribed gene in the mouse embryo reported to date belongs to a large heterogeneous class of genes termed **retroelements.** These are genes that encode unusual DNA polymerases that can synthesize DNA strands using RNA as a template instead of DNA. Such DNA polymerases are termed *reverse transcriptases* because the process of information flow from RNA to DNA is the reverse of most gene expression mechanisms (discussed in Chapter 2). The sequencing of the mouse genome has revealed more than one-third of the entire genome is related to retroelements. Many are thought to have entered the genome through infection by an ancient retrovirus, an RNA virus whose life cycle includes synthesis of DNA copies of its RNA genome that insert themselves into the cellular DNA after infection. New copies of viral RNA are thereafter synthesized from the newly acquired DNA sequences.

Over time, the retrovirus genes are silenced by cellular mechanisms that inhibit gene expression. The genomes of most animals, including humans, have evidence of repeated infection by retroviruses, followed by cellular control of virus expression. Many have lost large portions of the virus genes and have internal mutations that prevent gene expression. Most retroelements are silent and never expressed.

There are interesting exceptions, however, one of which is the recently reported early mouse embryo expression of a class of retroelements termed *murine endogenous retrovirus class L*, or MERV-L. As discussed in Chapter 4, expression of MERV-L was detected within 6 hours of fertilization and was shown to be important for progression of mouse embryos beyond two-cell stage. This observation is in general agreement with other experiments that have shown inhibition of reverse transcriptase activity also prevents mouse embryo development beyond two-cell stage.

This emerging body of information indicates that over time, mouse embryos have adapted to gene expression by ancient retroviruses and perhaps even adopted their gene products for their own use. It is important to discover the role of retroelement gene families, such as MERV-L, in maintaining pluripotency of embryonic cells and cultured stem cells.

Although it appears that the blastomeres do not require exogenous growth factors to proceed through the cell cycle, several lines of evidence suggest they may synthesize their own growth factors, a mechanism termed **autocrine stimulation**. One line of evidence is the need for the ras/raf pathway, a common signal transduction pathway that stimulates a critical cell cycle regulator, mitogen-activated protein (MAP) kinase, as described in Chapter 2. This pathway is activated by a variety of growth factors acting on receptors at the plasma membrane of somatic cells. A blockade of the ras/raf pathway blocks two-cell mouse embryos at the two-cell stage. Either this pathway operates differently in blastomeres from all other cell types or the blastomeres themselves secrete a growth factor that binds to its plasma membrane receptor and, in this way, activates the ras/raf pathway.

In addition to the detection of cyclin and Cdc2, these observations indicate that the cell cycle controls on blastomere cleavage are similar to somatic cells, with the exception that preexisting stockpiles of maternal mRNAs and proteins are deployed rather than de novo synthesis of essential components. Studies of the DNA replication and repair machinery operational in blastomeres are limited. Faithful DNA replication is the single most important job of a blastomere. Understanding blastomere-specific mechanisms of DNA replication is urgently needed, both for the production of human embryonic stem cells and for improving the safety and success of human IVF cycles.

Autocrine: stimulation of a cell process by a signal (e.g., a hormone) produced by the cell itself.

Summary

The activated egg undergoes its first cell division within approximately 24 hours. The resulting two-cell stage is the same mass as the egg, but each blastomere is half the size. It is for this reason that early cell divisions are termed *cleavages*. Five cleavage divisions are required to reduce the egg to the approximate size of somatic cells.

Blastomeres express stage-specific genes and have unusual energy requirements, such as pyruvate instead of glucose. Some genes are turned on in the zygote or two-cell stage and persist to the blastocyst stage. Other genes are turned on specifically for only one stage, such as the two-cell or four-cell stage, and not expressed at subsequent stages. An example of mouse two-cell stage-specific gene expression is histone deacetylase 2. Understanding the molecular basis for stage-specific deployment of specific genes reveals fundamentally important information about uncommitted cells and how they enter commitment pathways.

Early-stage blastomeres are totipotent, as evidenced by the ability of each blastomere of a two-cell embryo to give rise to an identical twin. But there is more recent evidence that some polarization of the mouse egg occurs before the first cleavage and that even by the two-cell stage, developmental fate of daughter cells has been influenced.

Progression of blastomeres through the cell cycle does not require the same growth factor stimulation as somatic cells, although it is possible that blastomeres secrete factors that stimulate their own cell cycle. The controls on replication of blastomere DNA may be different from adult somatic cells. Additional information is needed. Understanding how to support the cell cycle of single blastomeres could provide new approaches to the production of stem cell lines that bypass the present need for development to the blastocyst stage.

Review Questions (True or False)

1. Blastomeres are the large daughter cells of the cleaving egg.
2. Some genes are expressed specifically in discrete early cleavage stages and not throughout development to blastocysts.
3. Laboratory culture conditions to support the cell cycle of individual blastomeres to the stem cell stage are well known.
4. Although each two-cell stage blastomere of a mouse embryo can give rise to a new mouse (i.e., result in twins), recent evidence suggests they are not totally equivalent.
5. The cell cycle of early cleaving blastomeres has a short G1 phase, suggesting that cell components required for DNA synthesis are carried over from the previous cycle.
6. Of the 282 genes specifically expressed in the two-cell mouse embryo, only 35 were identified in GenBank when they were discovered in 2002.
7. The earliest gene transcribed in the mouse embryo belongs to a class of genes termed retroelements, whose function in the embryo is not known.

ADDITIONAL READINGS

Deb, K., Sivaguru, M., Yong, H. Y., and Roberts, R. M. (2006). Cdx2 gene expression and trophectoderm lineage specification in mouse embryos. *Science* 311: 992–996.

Gardner, R. L., and Davies, T. J. (2006). An investigation of the origin and significance of bilateral symmetry of the pronuclear zygote in the mouse. *Hum Reprod* 21: 492–502.

Hilragi, T., and Solter, D. (2004). First cleavage plane of the mouse egg is not predetermined but defined by the topology of the two apposing pronuclei. *Nature* 430: 360–364.

Katsuta, H. (1977). *Nutritional Requirements of Cultured Cells*. Baltimore: University Park Press.

Kigami, D., Minami, N., Takayama, H., and Imai, H. (2003). MuERV-L is one of the earliest transcribed genes in mouse one-cell embryos. *Biol Reprod* 68: 651–654.

Longo, F. J. (1987). *Fertilization*. London: Chapman and Hall.

Plusa, B., Hadjantonakis, A.-K., Gray, D., Piotrowska-Nitsche, K., Jedrusik, A., Papaioannou, V. E., Glover, D. M., Zernicka-Goetz, M. (2005). The first cleavage of the mouse zygote predicts the blastocyst axis. *Nature* 434: 391–395.

Slack, J. M. W. (1983). *From Egg to Embryo: Determinative Events in Early Development*. London: Cambridge University Press.

Strausberg, R. L., Feingold, E. A., Grouse, L. H., et al. (2002). Generation and initial analysis of more than 15,000 full-length human and mouse cDNA sequences. *Proc Natl Acad Sci USA* 99: 16899–16903.

CHAPTER 7

Early Nuclear Transfer Technology

Except in directions in which we can go too far there is no interest in going at all; and only those who will risk going too far can possibly find out just how far one can go.

T. S. Eliot

OVERVIEW

Early reproductive biologists studied amphibian eggs because their large size allowed visualization with existing microscopes, termed *dissecting microscopes*, whose magnifying power was limited. The story is told (see sidebar) that it was Hans Spemann's baby's hair that was delicate enough to assist in the isolation of individual nuclei in early cleaving newt eggs. As a result, Spemann reported in 1902 that embryonic salamander nuclei contained all the genetic information needed to guide the development of an entire salamander. This elegant observation contributed significantly to the concept that genetic information was encoded within DNA, the chemical structure of which would not be known for another 50 years.

Advances in microscopy, and the development of tools with finer control than a baby's hair, opened the way to ask questions about the genetic repertoire of more fully differentiated cells. A fundamentally important question was whether or not adult somatic cell nuclei had been genetically altered. Discovering the answer was important for many reasons. For example, if liver cells had lost genetic information or duplicated existing genetic information to function as liver cells, then cell functions were probably not reversible

Developmental Potential of Transplanted Nuclei

Hans Spemann contracted tuberculosis in 1899, just after Mendel's 1866 paper on genetics had been "rediscovered." As a result of the discovery of Mendel's work, it had become broadly appreciated among biologists that genetic traits were inherited in a somewhat predictable mathematical fashion. During his recovery in a sanatorium, Spemann familiarized himself with the developmental biology theories espoused by Weismann. At the time, it was understood that all individuals arose from a single fertilized egg that underwent multiple cleavages. Biologists were grappling with an explanation for cell differentiation—how did the profusion of different types of cells in each tissue happen? Weismann theorized that each time a cell divided, its chromosomes were split in half. Ultimately, the remaining chromosomes in a cell determined its function. Combining Mendel's and Weismann's theories meant believing that only sperm and eggs maintained the full complement of inherited traits. To Spemann's mind, this was a provocative possibility, but experimental data were lacking.

He lectured and studied at the University of Wurzburg while he worked toward his doctorate. He had chosen newts as his experimental system because the eggs were large and transparent and the laboratory conditions for their development were known. His first experiment (using a loop of hair from his newborn son) was to separate the blastomeres of a salamander two-cell embryo to determine whether one developed into the head and the other into the tail, as would be predicted by Weismann's theory. Instead, two complete salamanders developed. Ecstatic, he repeated the experiment by separating each blastomere of a four-cell embryo. Four salamanders developed. Then eight salamanders developed from each blastomere of an eight-cell embryo. However, after the eight-cell stage, development was erratic. He decided that at the 16-cell stage, the fates of each cell had been determined.

To test this theory, he coaxed both zygotic nuclei into one half of an egg and kept them there with another loop of his infant son's hair. After they had cleaved several times, he loosened the loop and manipulated one of the embryonic nuclei into the other side of the egg cytoplasm. This was the first nuclear transfer experiment. Amazingly, a second embryo formed. This proved that several nuclear divisions did not decrease the genetic information in the embryonic nucleus, which shot a big hole in Weismann's theory. Spemann's work was published in 1902.

Fifty years later, two groups of English scientists continued to explore the genetic repertoire of fully differentiated cells by transplanting nuclei into activated frog eggs. The first report was by King and Briggs in 1953. These investigators removed the genetic information from the eggs of the frog, *Rana pipiens,* and transferred in nuclei from cells of early *Rana* embryos and late *Rana* embryos. The early embryo nuclei directed the development to normal tadpoles in approximately one-third of the nuclear transplants, but no tadpoles were obtained from transplanted nuclei of later embryos. A few years later, in 1958, Gurdon, Elsdale, and Fischberg performed similar experiments with another frog, *Xenopus laevis*. Their experiments differed in that they did not remove the genetic information from the eggs before transplanting nuclei from *Xenopus* embryos because they had a genetic marker to distinguish offspring resulting from the transplants. These investigators reported development to adults after nuclei transferred from developmental stages as late as a pre-hatching tadpole. Thus, activated frog eggs were shown to remodel embryonic nuclei, but not adult cell nuclei, to give rise to new individuals. These were the first "nuclear transfer clones," although neither group of scientists used the term *clone* to describe the tadpoles and frogs that developed.

(termed *dedifferentiation*), and the process of cell differentiation into specific tissues involved specific genetic alterations. Alternatively, if liver cell nuclei possessed all the information for a new organism, then the fact they were liver cells was not due to genetic alterations but to epigenetic controls on gene expression. This information was fundamentally important to all of medicine, including developmental biology and cancer biology.

By the early 1940s, studies of frog embryo development had progressed to the discovery of laboratory methods for activation of frog eggs. This experimental system provided the opportunity to begin to explore the inherent developmental potential of nuclei from somatic cells.

Transfer of Nuclei Into Frog Eggs

Two groups of investigators (see sidebar) studied nuclei from embryonic and adult frog cells throughout the 1950s. Techniques had been developed to observe frog embryo development in the laboratory, from the moment of fertilization through the swimming tadpole stage and on to adult frogs. Techniques had also been developed to activate frog eggs without fertilization and to observe early stages of development in the laboratory. Cleavages occur rapidly after egg activation for removing the frog egg chromosomes (enucleation) and activating the frog egg, as described in Chapter 4. To assess the potential for chromosomes in the nuclei of various types of frog cells to direct development, the nuclei were transplanted individually into enucleated frog eggs. The results of these studies indicated that embryonic nuclei, but not nuclei from adult somatic cells, could direct development of new frogs. Following the results of those experiments, by the early 1960s medical students and developmental biologists were being taught that embryonic nuclei retained all the information needed to direct the development of new frogs, but adult cell nuclei had been permanently altered and could no longer direct embryonic development.

The belief that the genetic information in adult somatic cells was permanently altered shaped scientific thinking for at least two decades. Similarities between the growth of embryonic tissues and the growth of cancerous tissues were obvious, but the concept that normal cells could dedifferentiate into cancerous cells was discounted because of the evidence that normal adult cell nuclei could not be reprogrammed to support embryonic development. Nonetheless, throughout this period studies of DNA had revealed its basic structure, and numerous studies by many research teams of the DNA content of many types of cells revealed no gross differences between embryonic and adult cells with respect to chromosome number, DNA content, or DNA sequences. Other explanations for the inability of adult cell nuclei to support embryonic development were sought.

The frog egg experiments had demonstrated several important egg functions, including the fact that eggs did not need to be activated by sperm to remodel somatic cell nuclei. Remodeling chromatin is clearly what eggs are designed to do once they are activated, as described in Chapter 3. Moreover, nuclear remodeling and subsequent cleavages occur with or without the presence of the egg chromosomes. Thus, the cellular components involved in nuclear remodeling and early embryonic development preexist in some form in the frog egg cytoplasm.

Given the wide variety of functions cells must perform, it seemed likely that some nuclei could be more easily reprogrammed than others. An important constraint on nuclear reprogramming by eggs is that it must take place simultaneously with egg activation to set up sustained cleavages. As described in Chapter 3, metaphase II–arrested eggs are in a carefully controlled cell cycle arrest. That arrest must be overcome to allow nuclear remodeling and cleavage. But the egg has a finite lifetime. Aged eggs are more readily activated but fail to sustain as many cleavages as do "young" metaphase II eggs. Thus, the narrow time interval to accomplish nuclear remodeling and initiate successful cleavages may be insufficient for reprogramming some somatic cells. One possible solution to this problem was to allow the nucleus of a somatic cell to be remodeled in stages.

In 1970, Gurdon and Laskey tried this approach. They reported the development to adult frogs following a two-step transfer of nuclei originally taken from a monolayer of cultured epithelial frog cells (Figure 7.1). The epithelial cell nucleus was first transferred into an activated egg, which then cleaved to an early embryo. A nucleus from the early embryo was then transferred into another activated egg from which the genetic material had been removed. A small percentage (less than 5%) developed through the tadpole stage to adult frogs. These results demonstrated that skin cell nuclei maintain the full genetic potential of an individual. The problem is in the nuclear remodeling.

FIGURE 7.1

Schematic of two-stage
nuclear transfer experiments
in frogs.

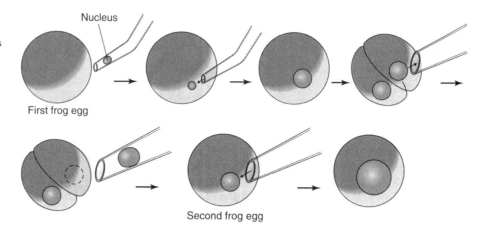

Nucleus

First frog egg

Second frog egg

Once Gurdon had established that frog eggs could remodel nuclei from adult frog cells, it was obvious to try to understand what would happen if nuclei from other species were transferred into frog eggs. The first cross-species human–frog nuclear transfers were reported in the late 1970s. The human nuclei were recovered from HeLa cells, a human cell line established in culture from a cervical carcinoma. HeLa cells have been used in many laboratories for a wide variety of experiments for several decades. They were an obvious choice for the experiments designed to see the effect of frog egg cytoplasm on gene expression by human cells.

Gene Expression: Transcription and Translation

The sensational human–frog experiments are actually built on other work, a description that benefits from a brief review of the basic steps of gene expression. As discussed in Chapter 2, genes are unique, and specific polymers of four molecules distinguished by side groups of adenosine (A), cytosine (C), guanosine (G), and thymidine (T). The arrangement of the molecules in the polymer determines the nature of the protein specified by the gene. Genes are joined end to end in the long strands of DNA that make up the chromosomes. Genes are highly variable in length and are distinguished from each other by specific "start" and "stop" sequences (see Figure 2.9).

The expression of genes into new proteins requires a series of well-coordinated steps (Figure 7.2). First, the specific base sequence in one strand is transcribed into single-stranded RNA. Enzymes, termed *RNA polymerases*, synthesize the RNA to uniquely mimic the DNA sequences in the gene. The RNA strands contain molecules with three of the same four side groups (A, C, and G) and uridine (U) instead of thymidine. Three types of RNA are **transcribed** from individual genes: messenger RNA (mRNA), ribosomal RNA (rRNA), and transfer RNA (tRNA).

Transcription: the process of synthesizing messenger RNA (mRNA) from the gene sequence as the template.

FIGURE 7.2
Flow scheme of transcription
and translation.

Gene Expression: Transcription and Translation **107**

mRNA is the template for protein synthesis. rRNA is a structural molecule that combines with specialized proteins to form the ribosome, a globular structure fundamentally important to protein synthesis. tRNA is a small highly structured molecule that recognizes and transports the specific amino acids needed for the growing protein chain. Together, the RNAs constitute an efficient assembly line factory that produces proteins. The process is remarkably complicated and well controlled.

A seemingly boundless repertoire of signals is involved in turning off and turning on RNA synthesis from genes. Hormones, growth factors, invasion by bacteria or viruses, starvation, gluttony, and sunshine are examples of stimuli for gene expression. Each stimulus excites an intracellular signal transduction pathway that, principally through a series of phosphorylations and dephosphorylations (the domino effect described in Chapter 2), in turn mobilize transcription factors to turn on or off gene expression in the nucleus. Some transcription factors are proteins that bind specifically to an area of the chromosome and expose the gene to the RNA polymerases. In addition to RNA polymerases, transcription involves a large complex of enzymes and other cofactors that bind to a specific region of the gene, termed the *promoter*, and systematically manufacture the RNAs, which then leave the nucleus for the cytoplasm. Multiple RNAs can be transcribed from the same gene at the same time. rRNAs combine with ribosomal proteins in the cytoplasm to form the tightly coiled structure that forms the physical platform for protein synthesis. The tRNAs coil up and remain free spirits to search the cytoplasm for their specific amino acid.

As detailed in Chapter 2, mRNA slides between the two subunits of the ribosomes (Figure 7.2), exposing its sequence in groups of three molecules, the triplet codon, which defines an individual animo acid. The process of reading the mRNA sequence and synthesizing the corresponding protein is termed *translation*.

Proteins are synthesized at the rate of 20 amino acids per second. Thus, one molecule of insulin can be synthesized in approximately 3 seconds. Each mRNA can simultaneously direct the synthesis of multiple growing protein strands, and each gene can transcribe multiple mRNAs simultaneously. Thus, millions of copies of proteins can be produced from one gene sequence in a matter of minutes, and production can be halted just as fast. The speed, precision, and responsiveness of gene expression are awe-inspiring cellular assets.

In no cell is it more finely tuned than in the egg. As described in Chapters 3 and 4, unlike dividing somatic cells that synthesize and degrade proteins, as directed by either cell cycle machinery or specific cellular functions, the egg synthesizes large stockpiles that must be deployed as needed. Ribosomes and ribosomal precursors are stockpiled in massive quantities by the egg. Large stores of mRNAs, termed *maternal messages* in eggs, such as cMos, which is essential for metaphase II arrest (see sidebar, Chapter 3), are also stockpiled. Other cellular components essential for rapid and accurate early cleavages are stored as proteins, such as the DNA polymerase detected in relatively high concentration in metaphase II–arrested mouse eggs. The systematic deployment of these stockpiles for the multiple cycles of DNA replication and cleavage is a property of eggs that remains poorly understood, especially with respect to human eggs.

Transcription and Translation in Frog Eggs

The nuclear transfer experiments conducted in the 1950s led to an appreciation of the frog egg as an experimental system with which to explore controls on gene expression. Frog eggs can be recovered without damage to the frog (they can be hugged out by a belly squeeze, mimicking the bull frog) and thus represent an unlimited supply of eggs for study.

Injection of specific mRNAs into frog eggs, such as the mRNA for human hemoglobin, resulted in the synthesis of hemoglobin proteins in the frog egg cytoplasm (Figure 7.3). In addition, injection of genes in the form of DNA fragments were transcribed into messages and then translated into proteins. Several groups of investigators studied the transcription and translation of genes and mRNAs injected into frog eggs. The results of their studies led to valuable new insights into the controls on the processes of transcription and translation.

One important control is the need to add a string of "A"s at the beginning of the mRNA to signal that it is prepared for translation. This process is termed **polyadenylation**, and it is an important control of gene expression. mRNAs that are not polyadenylated may thus be stockpiled and not translated into protein. Therefore, one way to recruit specific maternal messages from the egg's stockpile is to bring about polyadenylation. This is an important egg control mechanism for both egg activation and early cleavages.

Polyadenylation: the process of adding a series (sometimes hundreds) of adenine nucleotides to the end of an mRNA molecule to signal its binding to ribosomes and translation into a protein.

Protein stockpiles may also have specific control mechanisms. After translation, many proteins require additional modifications before they become functional. Such modifications include splitting off a small segment that inhibits activity or the need to add side chains, such as fatty acids or carbohydrates, to allow the protein to be transported to its point of action. For example, insulin is synthesized as a long precursor protein that is cleaved into two polymers that then interact with each other in a specific fashion to bring about folding of the molecule to fit precisely into the

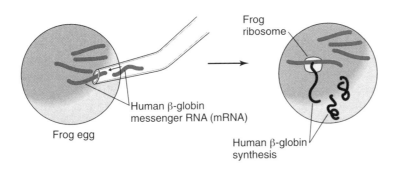

Frog ribosome

Human β-globin messenger RNA (mRNA)

Frog egg

Human β-globin synthesis

FIGURE 7.3
Injection of human mRNAs into frog eggs.

insulin receptor. Precursor proteins are not active and thus can be held in reserve until needed.

These considerations highlight the many possible mechanisms available to an egg both to stockpile cellular components and to control their deployment. This is in sharp contrast to somatic cells whose gene expression, both on and off, is more tightly controlled by environmental stimuli. Size differences alone between the germinal vesicle and somatic nuclei suggest different mechanisms of control of gene expression. Somatic cell chromatin must be opened up by modifications to the chromatin or by transcription factors to allow transcription to occur, whereas the huge scaffolding available to egg chromatin in the germinal vesicle may obviate the need for transcription factors, as discussed in Chapter 3. Transferring a nucleus from a somatic cell into an egg is therefore not simply transferring a bag of genes. The somatic cell genes have been under an entirely different set of regulations that must be set aside to allow the egg cytoplasm to take over control of DNA synthesis and initiate cell cleavage. This process has been termed **nuclear remodeling**.

Nuclear remodeling: changing a somatic cell nucleus into the large nucleus of a zygote or cleaving blastomere.

Reprogramming a Nucleus

Genes can be broadly divided into three groups: housekeeping genes, cell cycle genes, and cell function genes. Housekeeping genes encode proteins that contribute to routine cell functions, such as respiration, protein synthesis, protein degradation, and DNA repair. Many of these genes are highly conserved between species, some even between plants and animals. Some are expressed at all times, termed *constitutive expression*, but others are expressed only at specific times, such as heat shock proteins, which appear with fever. It is presumed that, for the most part, housekeeping genes may not need to be reprogrammed to allow the nucleus from a differentiated cell to direct embryonic development.

Cell cycle genes were discussed in Chapter 2 (see Table 2.1). Although many aspects of blastomere cell cycles are similar to somatic cells, there are distinct differences, especially related to the foreshortened G1 phase. So, although cyclin synthesis and degradation appear to be highly conserved between species and between embryonic and somatic cells, the details of cell cycle regulation in a somatic cell need to be reprogrammed for successful and sustained blastomere cleavage.

Cell function genes are the essence of a differentiated cell. Expression of specific sets of genes defines cell types. Some cells express genes not expressed in any other types of cells. For example, the proteins that compose the lens of the eye are not expressed by other cells. The zona pellucida proteins that provide the protective coating for eggs are not expressed by other cells. Hemoglobin, the protein that carries oxygen to all cells of the body, is expressed specifically in red blood cells. Cell surface receptors for hormones, such as follicle-stimulating hormone (see Chapter 3), are expressed by only a few cells, principally those in the ovary and the testis. The immune system is redundant with examples of cell surface receptors that enable a specific immune responsive cell to mount an attack against an infectious agent.

These examples highlight the nearly incomprehensible spectrum of cellular behavior encoded by the human genome. In contrast, do the egg and cleaving blastomeres express fewer genes and in this way avoid such highly differentiated cell functions? The egg itself is a highly differentiated cell, as is the sperm. So is it the combination of two new genomes that avoids differentiation into committed cell types for the first several cleavage divisions? These are the questions that have plagued developmental biologists for decades, and their answers will provide valuable information about mechanisms necessary to bring about nuclear reprogramming.

What is clear, with respect to nuclear reprogramming, is that the highly specialized cell functions need to be turned off, and the nuclear chromatin needs to be prepared to undergo S phase nearly as soon as the nucleus is injected into the egg. This is undoubtedly a complicated requirement.

It is important to note that, as discussed in Chapters 5 and 6, more than half of the mRNAs recovered from mouse eggs and early embryos are from unknown genes in GenBank. This emphasizes the limitations to our understanding of the basic cellular processes that bring about reprogramming of both sperm and egg gene expression to give rise to totipotent embryonic cells, followed by several cycles of DNA replication and cell cleavage in preparation for commitment. The same basic reprogramming requirements may be needed for somatic cell chromosomes to direct egg cleavage into pluripotent stem cells.

As discussed previously, HeLa cell nuclei were injected into frog eggs in 1974 in the Gurdon laboratory. Human mRNA and proteins were detectable for several days after the nuclear transfer. This landmark experiment demonstrated several fundamentally important findings:

1. Frog eggs do not contain enzyme systems that specifically degrade foreign DNA, as do many bacteria. This was a surprising finding given the importance of maintaining the species specific integrity of the genome. The lack of an inherent mechanism for destroying foreign DNA seems especially curious for an animal whose eggs are fertilized among the many foreign species present in pond water. This observation is also true for eggs from other species, a finding supported by Ben Brackett's report that rabbit sperm could carry radioactive viral DNA (SV40) into rabbit eggs at fertilization. These two observations are consistent with the lack of detectable DNA-degrading enzymes (deoxyribonucleases) in mouse eggs. Zygotes also may not have a mechanism for destroying foreign DNA given the ease with which transgenic mice are produced (see sidebar, Chapter 5).

2. Frog transcription factors and RNA polymerases can transcribe at least a subset of human genes, supporting the highly conserved nature of at least some gene expression controls across several animal phyla. Numerous experiments injecting human nuclei into frog eggs followed these early experiments, with multiple reports of expression of specific human genes.

Studies to specifically reveal human genes that cannot be expressed in frog eggs are lacking.

3. Frog ribosomes, translation factors, and tRNAs can recognize and synthesize human proteins from the human mRNAs, providing further support for the highly conserved nature of mechanisms of protein synthesis.

In summary, although the frog egg experiments yielded enormous information about egg controls on gene expression, the transfer of the HeLa cell nuclei did not lead to cleavage. Determining whether or not this was due to an incompatibility between frogs and humans or whether mammalian somatic cell nuclei were fundamentally different from frog somatic cell nuclei requires the development of nuclear transfer techniques for work with mammalian eggs, which is discussed in Chapter 10.

Once the frog skin cell experiments described in this chapter had been reported, scientists attempted similar experiments with rabbit (Bromhall, 1975) and mouse (Modlinski, 1978) eggs. Advances in microscopy manipulation tools, termed *micromanipulators*, improved the ability to stabilize and inject mammalian eggs with nuclei directly into the cytoplasm. These investigators transplanted nuclei from early rabbit and mouse embryos, respectively, into fertilized zygotes. In each series of experiments, the zygotes that survived the micromanipulation successfully remodeled the injected nuclei, but only a few cleavage divisions occurred. Developmental arrest was attributed to the fact that the nuclear transfer eggs were tetraploid. They contained both the zygote chromosomes and the chromosomes contributed by the transplanted nuclei.

During the late 1970s and early 1980s, Karl Illmensee worked in the laboratory of Peter Hoppe at the Jackson Laboratories in Bar Harbor, Maine. In 1981, they reported a series of experiments designed to address the concerns about tetraploidy. They transplanted nuclei from mouse blastocyst cells into mouse zygotes whose pronuclei were physically removed by micropipets (Figure 7.4). Thus the resulting reconstructed zygotes contained only the chromosomes from the transplanted nuclei and were diploid, not tetraploid. The results of their experiments, described in the following section, revolutionized thinking about the developmental potential of differentiated nuclei. However, the inability of other investigators to repeat their work cast a cloud on the integrity of their report, which remains today, although their basic findings have now been repeated by other research teams.

The 1981 report by Illmensee and Hoppe addressed two experimental goals: One was to determine whether nuclei from trophoblast cells had the same ability to support embryonic development as nuclei from *inner cell mass* (ICM) cells, and the second was to determine the potential of the nuclei from cells of the blastocyst to direct the development of a new mouse. They reported that nuclei from trophoblast cells rarely supported development beyond the early cleavage stage: 179 eggs were manipulated, 68 survived both the transfer of trophectoderm nuclei and the removal of the two zygote pronuclei, 34 cleaved at least once, 6 developed to morulae, and 1 developed to a blastocyst (1.5% of surviving eggs).

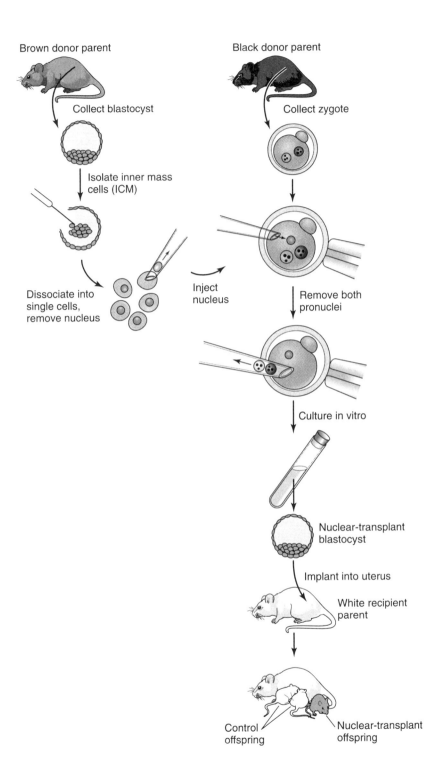

Brown donor parent

Collect blastocyst

Isolate inner mass cells (ICM)

Dissociate into single cells, remove nucleus

Inject nucleus

Black donor parent

Collect zygote

Remove both pronuclei

Culture in vitro

Nuclear-transplant blastocyst

Implant into uterus

White recipient parent

Control offspring

Nuclear-transplant offspring

In contrast, nuclei from ICM cells supported a 10-fold higher rate of development to blastocysts: 363 eggs were manipulated, 142 survived, 96 cleaved at least once, 25 developed to morulae, and 23 developed to blastocysts (16% of surviving eggs). Moreover, a landmark result was the birth of offspring after transfer of cleaving embryos from transplanted ICM nuclei into the uteri of pregnant mice (Figure 7.4). Thus, in a manner analogous to the frog experiments, mouse ICM cell nuclei could be reprogrammed by zygote stage eggs to direct the development of new individuals. Although the authors did not use the term *clone* in their report, these experiments actually produced the first cloned mice.

In similar experiments conducted the same year, J. Modlinski, working at the Research Council in London, also found that ICM cell nuclei more efficiently supported development to blastocysts after transfer into mouse zygotes than did nuclei from trophoblast cells. Unlike the experiments of Illmensee, however, Modlinski did not remove the pronuclei from the zygotes, illustrating that whether or not the resulting reconstructed embryos were diploid or tetraploid, ICM cell nuclei more efficiently supported cleavage than trophoblast cell nuclei.

Transplantation of Embryonic Nuclei Into Fertilized Eggs

The results from amphibian systems suggested that adult cell nuclei could dedifferentiate and fully support embryonic development. The reports of Illmensee and Hoppe and Modlinski emphasized the high death rate for zygotes whose nuclei were removed and replaced with a nucleus from another cell. Therefore, throughout the 1980s research teams sought new techniques. Because mammalian eggs are so much smaller than amphibian eggs and blastomere nuclei are so much larger than somatic cell nuclei, the techniques of nuclear transplantation into mammalian eggs had to be perfected.

To avoid puncturing the plasma membrane of the mammalian egg, Davor Solter and his colleagues at the Wistar Institute in Pennsylvania took advantage of the property of some proteins, most notably those expressed on the surface of Sendai virus, to cause two plasma membranes to fuse together. In 1983, they reported a series of experiments in which they removed both pronuclei from zygotes and replaced them with pronuclei from another zygote. The two pronuclei were removed at the same time with a glass pipette similar to the one described by Illmensee, and new pronuclei were introduced into the enucleated zygote by a membrane fusion technique (Figure 7.5). The pronuclei to be transplanted into the egg were drawn up into the glass micropipette, exposed to a small amount of inactivated Sendai virus, and then placed into the **perivitelline space** between the zona pellucida and the zygote plasma membrane (Figure 7.5). The manipulated zygote was returned to the incubator and fusion took place within approximately 1 hour. The developmental potential of the reconstructed zygotes was almost the same as nonmanipulated controls.

Perivitelline space: the area between the egg plasma membrane and the zona pellucida.

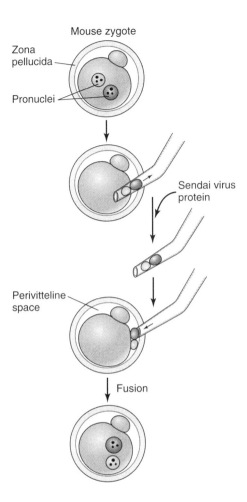

FIGURE 7.5
Transplantation of mouse pronuclei by membrane fusion.

Mouse zygote

Zona
pellucida

Pronuclei

Sendai virus
protein

Perivitteline
space

Fusion

Using this technique, McGrath and Solter demonstrated that nuclei from cleaving blastomeres could also be fused to enucleated zygotes by inactivated Sendai virus (Figure 7.5). In 1984, they published a report that refuted the work of Illmensee and Hoppe in 1981. Their studies demonstrated that 99% of zygotes could survive removal of both pronuclei. As before, if fused with pronuclei from another zygote, 95% of the reconstructed zygotes developed to blastocysts in culture. If fused with the nucleus from a two-cell embryo, 81% cleaved once or twice, but only 15% of those developed to blastocysts. If fused with the nucleus from a four-cell embryo, 95% cleaved once or twice, but none developed to a blastocyst. If fused with the nucleus from a morula, all cleaved once or twice, but none developed to blastocysts. Finally, if fused with an ICM cell, an experiment analogous to that of Illmensee and Modlinski, all reconstructed zygotes cleaved once or twice, but none developed to blastocysts.

These results were so different from those reported in 1981 by the other investigators that McGrath and Solter repeated a series of experiments in which nuclei were injected directly into the zygote cytoplasm instead of using their fusion technique. The outcome did not change. None of the enucleated zygotes injected with ICM nuclei (72 of 231 zygotes manipulated survived the reconstruction procedure) developed past one or two cleavages, again in disagreement with the reports of Illmensee and Modlinski.

Summary

By the 1950s it was known that DNA was the molecule responsible for genetic information and that it was contained within the nucleus of the cell. Embryonic nuclei (e.g., those in the newt) contained all the information needed to guide the development of an adult newt.

Interest turned to controls on cell differentiation. A fundamentally important question was had the DNA in an adult cell (e.g., a liver cell) been permanently altered so it would function only as a liver cell? Many experiments in several laboratories eventually established the concept that adult frog cell DNA was not permanently altered in gene sequences but was under epigenetic control, and in principle the DNA from any cell injected into an egg could direct the development of a new frog, if the epigenetic controls could be reversed, termed *reprogramming*.

By the mid-1980s, frog eggs were a well-established experimental system for studies of molecular controls on gene expression, including protein synthesis. Genes could be introduced into the frog eggs in the form of purified DNA or contained within the nuclei of cells under study. Under certain circumstances, frog eggs could remodel the chromosomes in the nuclei of some frog somatic cells, leading to successful embryonic development to tadpoles and, in some cases, to adult frogs.

The situation was less clear for mammalian eggs. Some chromosome remodeling occurred when somatic cell nuclei were injected or fused to mouse eggs, but embryonic and fetal development was lacking, perhaps impossible, despite the increasing body of information that somatic cells retained all genetic information throughout differentiation. Genes can be classified into three rough categories: housekeeping, cell cycling, and cell function. To reprogram a somatic cell, the housekeeping genes may not require reprogramming, but the genes controlling the cell cycle and specific cell functions do require reprogramming.

By the 1990s, the importance to nuclear remodeling of some form of synchrony between the injected nucleus and the cytoplasm of the egg was appreciated. But the prevailing view was there was a fundamental difference between amphibians and mammals. Although much has been learned in the last 100 years, it is still sobering to realize that over half of the genes in eggs and early embryos are still not characterized.

Review Questions (True or False)

1. The first nuclear transfer experiment was performed over 100 years ago.
2. Once a cell has differentiated into an adult cell, it can never regain pluripotency.
3. It is easier to remodel an early-stage embryonic cell than an adult somatic cell.
4. Adding adenine to mRNA (polyadenylation) keeps it from being expressed.
5. Frog eggs cannot transcribe foreign DNA.
6. Proteins are synthesized on the ribosomes at a rate of 20 amino acids per second.
7. Remodeling a somatic cell can be enhanced by passing it through two eggs.
8. Every time a cell divides, it loses half its genes.

ADDITIONAL READINGS

Davidson, E. H. (1986). *Gene Activity in Early Development*. London: Academic Press.

Gurdon, J. B., Elsdale, T. R., and Fischberg, M.(1958). Sexually mature individuals of *Xenopus laevis* from the transplantation of single somatic nuclei. *Nature* 182: 64–65.

Gurdon, J. B., Laskey, R. A., and Reeves, O. R. (1975). The developmental capacity of nuclei transplanted from ketarinized skin cells of adult frogs. *Embryol Exper Morphol* 34: 93–112.

Modlinski, J. A. (1981). The fate of inner cell mass and trophectoderm nuclei transplanted to fertilized mouse eggs. *Nature* 292: 342–343.

McGrath, J., and Solter, D. (1986). Nucleocytoplasmic interactions in the mouse embryo. *J Embryol Exper Morphol* 97: 277–289.

Solter, D., Aronson, J., Gilbert, S. F., and McGrath J. (1985). Nuclear transfer in mouse embryos: activation of the embryonic genome. *Cold Spring Harbor Symp Quant Biol* 50: 45–50.

Tsunoda, Y., Yasui, T., Shioda, Y., Nakamura, K., Uchida, T., and Sugie, T. (1987). Full-term development of mouse blastomere nuclei transplanted into enucleated two-cell embryos. *J Exper Zoo* 242: 147–151.

Willadsen, S. M. (1986). Nuclear transplantation in sheep embryos. *Nature* 320: 63–65.

PART

III

Embryonic Stem Cells

The Blastocyst and Inner Cell Mass Cells

Indeed, each blastomere has the potential to form any cell of the body.

James Thomson

OVERVIEW

As described in Chapter 6, the geometry of the cleaving blastomeres changes with each cell division. Four-cell blastomeres are all the same size, but six-cell blastomeres are two different sizes: four are smaller (eight-cell size) and two are larger (four-cell size) (Figure 8.1). Thus, only at the 4-, 8-, and 16-cell stages are all the blastomeres the same size. Because of the spacial constraints within the zona pellucida, the blastomeres rearrange themselves with each cleavage event. Each of the eight-cell stage (early morula) blastomeres has contact with other blastomeres and the outside of the cell cluster. However, by the 16-cell stage, one or two of the blastomeres is completely trapped inside the cell mass and has no contact with the outside world (Figure 8.1).

The enclosure of a cell, or a couple of cells, inside the morula signals the first embryonic commitment event: The cells on the outside become committed to giving rise to the placenta; those trapped inside give rise to the embryo. The outside cells are termed *trophoblast* and give rise to trophectoderm, the foundation of the placenta; the inside cells are termed **inner cell mass (ICM)** cells. Under some circumstances, ICM cells can contribute to the trophoblast layer, but trophoblast cells are not thought to contribute to ICM. Best studied in the mouse, there is evidence to embrace similar commitment events in

ICM: inner cell mass, the cluster of cells at one end of the blastocyst that gives rise to stem cells.

FIGURE 8.1
Six-cell stage blastomeres are unequal in size.

Four-cell   Six-cell   Eight-cell (morula)   16-cell (morula)

the rabbit, cow, sheep, and presumably human. The precise cellular mechanisms that dictate this first commitment event remain elusive. The importance of fully understanding this critical developmental event cannot be overstated.

The decision to become either an ICM or trophoblast cell is characteristic of the type of commitment decisions that occur throughout embryonic development. One of two paths is chosen, and once chosen, as is the case with trophoblast, back differentiation does not seem to occur. Only those cells that have not taken the next step of differentiation remain available for the full spectrum of future differentiation events. These are also the basic properties of stem cells. It is speculated that specific sets of genes are turned off and new sets turned on at each differentiation event, but evidence for this is incomplete. This is an active area of investigation, with new information reported regularly. What is known is that the ICM cells and the trophoblast cells develop distinct characteristics with respect to cell cycle times and gene expression. The embryonic genome is clearly expressed by this stage, but some important egg functions persist as well.

Curiously, laboratory conditions for culturing individual blastomeres have not been developed. Alone, the blastomeres do not continue to cleave. It is possible that, because of their large size, they need physical and nutrient support for their cell cycles. In addition, progression through their cell cycles may require stimulation by factors they secrete themselves. Termed *autocrine stimulation*, described in Chapter 6, this is a well-known mechanism for several cell functions. Such factors, usually proteins, are synthesized in the cytoplasm and secreted through the plasma membrane. Once outside the cell, they bind to specific receptors on the surface of the plasma membrane that are primed to interact with signaling molecules inside the cell. The signaling molecules bring about transitions in the cell cycle (described in Chapter 2). Although speculative, there is evidence that such factors are produced and are important to blastomere cleavage. It is possible that one isolated blastomere may not synthesize factors in sufficient concentration to activate receptors and promote progression through the cell cycle. Until this problem is solved, the fate of blastomeres that do not experience either an inside or outside

121

position in the morula will not be known. In addition to providing fundamentally important information about early developmental commitment events, developing the technology to culture individual blastomeres could greatly enhance the efficiency of deriving stem cells from activated eggs, as discussed in Chapter 10.

Young Chung and his colleagues at the laboratories of Advanced Cell Technology have recently reported that single blastomeres from mouse eight-cell embryos continue to cleave if surrounded by mouse embryonic stem (ES) cells. Forty-four single blastomeres cultured individually failed to cleave, whereas 36 of 175 blastomeres (21%) wrapped in mouse ES cells continued to cleave, and 12 (7%) gave rise to continually dividing cell lines. This supports the notion of autocrine stimulation, but the failure of 80% of the blastomeres to continue cleaving in the presence of ES cells suggests that it may be blastomere-specific autocrine stimulation, perhaps requiring cell-to-cell contact, as has been suggested for equivalence groups.

Equivalence Groups

Studies reported in 1979 and 1980 of the simple microscopic flatworm, *Caenorhabditis elegans*, revealed that from a group of six cells that were precursors to its egg-laying structure, only three cells gave rise to it. The other cells then went on to contribute to primitive skin. However, ablation of the three cells initially committed to the egg-laying structure caused the remaining three to abort their skin destination and to give rise to the egg-laying structure instead. This showed that the six cells were multipotential and could compensate for each other if necessary, thus giving rise to the concept of an equivalence group of cells. This is an attractive theory that carries with it the caveat that equivalence groups have a primary (default) fate and a secondary fate that is assumed only if not needed for the default. Once the secondary fate is assumed, however, the primary fate cannot be reassumed, thus establishing a time frame within which decisions must be made. In this way, the direction of cell commitment events is also established. This concept seems fundamentally important and redundant throughout embryonic development, as discussed further in Chapter 9.

The Blastocyst

As described in Chapter 6, the transition from cleaving blastomeres to the ball of cells termed the *morula* happens at approximately the same time that the embryo leaves the oviduct and enters the uterus. It is as if the oviduct provides an environment protected from outside influences, such as serum growth factors. In this way, the oviductal environment allows the cleaving egg to accomplish several replications of DNA, finally achieving a cell size more in keeping with somatic cells. As discussed in Chapter 2, dividing somatic cells must grow in size for each

FIGURE 8.2(a)
Compacted morula.

cell division. Factors that stimulate specific cell pathways appear to be necessary for cell growth, as described in Chapter 6. Because continued cell division is essential for continued development, the embryo must leave the protected environment of the oviduct and enter the growth factor–rich environment of the uterus.

Several things happen once the cells become committed due to their position in the morula. The outer cells develop tight junctions with each other, termed **compaction** (Figure 8.2a). Compaction has been most thoroughly studied in the mouse because it occurs by the 12- to 16-cell stage. Compaction does not occur in other species until somewhat later stages, the 16- to 32-cell stage in the cow, goat, and possibly human. The individual blastomeres flatten against each other to form a cohesive cluster that loses the appearance of a group of individual cells. The expression of a specific membrane ion channel enables the outer cells to actually pump water and nutrients in the direction of their interior surfaces. In this way, the cells become polarized. Their physical shape changes; they become elongated and take on the appearance of flattened epithelial cells.

As the cells pump water and ions inward, the fluid collects together on one side of the interior (Figure 8.2b), giving rise to the **blastocyst**. The fluid-filled

Blastocyst: forms the ICM and an outer layer of trophoblast cells sealed together to form a trophectoderm that maintains the fluid-filled blastocoel.

FIGURE 8.2(b)
Early cavitating mouse blastocyst.

chamber is the **blastocoeloel** cavity. It is thought to help provide necessary nutrients for the expanding ICM. The trophoblast cells opposite the ICM (termed the **abembryonic** end of the blastocyst) cease to divide and begin to accumulate lipid deposits. They also undergo **endoreduplication**, a process by which the DNA is replicated many times without cell division. The value of this is unclear but may be related to the need to maintain a sealed cavity because the process of mitosis, of necessity, must break the gap junctions that constitute the seal between the trophoblast cells. Endoreduplication therefore provides a mechanism for increasing the copy number of genes, and hence gene expression, without the process of mitosis, which would break the seal.

Of the 9,718 unique genes detected in preimplantation mouse embryos by Ko and his colleagues, described in Chapter 6, 4% (387) were expressed specifically in morulae and 1.7% (161) were expressed specifically in blastocysts (Table 8.1). In keeping with the observed single-stage expression characteristics of 22% of the unique genes, only 12 genes were turned on in morulae and still detected in

TABLE 8.1	Genes Expressed Specifically in Mouse Blastocysts	
A00537	T haplotype-specific element	
A00913	Topoisomerase-related protein	
A01176	Homolog to rat CIC-K1 protein	
A00893	Connexin 31	

blastocysts. Thus, most genes turned on exclusively in morula were turned off in the blastocyst, in keeping with the notion that progression to the next stage of development requires gene silencing in addition to gene expression.

It is important to note that expression of some genes involved in the formation of the blastocoel, such as the ion channel that is responsible for the transport of water, is not dependent on the number of cell divisions that have occurred but on the length of time after activation of the egg. Some mouse blastomeres that undergo cell cycle arrest are nonetheless capable of pumping water into vacuoles within their own cytoplasm approximately 94 hours after egg activation, the time at which the blastocoel cavity first appears in developing mouse embryos. This is another example of the cascade of cellular reactions that results from the systematic deployment of components stored in the egg's cytoplasm. This observation makes it all the more curious that ICM cells do not also express the ion channel at the same time it is expressed by trophoblast cells. One possible explanation for the difference is the existence of both positive and negative controls on deployment of egg stores, some being deployed by "outer" cells and others by "inner" cells. This line of reasoning indicates that maternal cytoplasmic stores are in some way responsible for activation and control of the embryonic genome. Nuclear transfer studies described in Chapter 10 explore this possibility in more detail.

The Inner Cell Mass

Clustered at one end of the blastocyst, the ICM is the first equivalence group that ultimately gives rise to the embryo. Protected from outside influences by the surrounding trophoblast cells, the ICM continues to grow in cell number, closely associated with each other. Efforts to understand cellular controls on ICM development have included attempts to discover all the genes expressed by trophoblast cells and ICM cells. Some researchers have detected and characterized individual genes, whereas others have sought to detect and identify all genes expressed by each cell with the aid of high-throughput analysis methods. James Adjaye and colleagues at the Max Planck Institute in Germany detected 7,481 genes expressed by human blastocysts, of which only 292 were expressed exclusively in the ICM and 345 expressed exclusively in trophoblast. Their work confirmed and extended

TABLE 8.2	Genes Differentially Expressed in Blastocyst Cells
ICM Cells	**Trophoblast Cells**
SOX2	hCG
Oct4	KRT18
REX1	HANDI
NANOG	PSG3
DAB2	CDX2
DNMT3B	TBX1

studies from numerous other laboratories that revealed differential gene expression at the blastocyst stage, as listed in Table 8.2.

When the ICM of the mouse comprises approximately 50 cells and the entire blastocyst contains approximately 150 cells, a second differentiation event occurs. The layer of cells that line the blastocoel side of the ICM take on a new flattened appearance and form the primitive endoderm, the layer of cells that contribute to the **extraembryonic membranes**, including the **amnion**, the fluid-filled sac that houses the developing fetus. This second differentiation event occurs at approximately the same time the mouse blastocyst sheds its zona pellucida, a process called **hatching**. Once outside the zona pellucida, the blastocyst can attach itself to the lining of the uterus. The interaction between the blastocyst and the uterine lining is termed *nidation* and heralds the process of **implantation**. This therefore marks the end of the preimplantation period of development.

Implantation: the process of trophoblast attachment and outgrowth into the lining of the uterus.

Implantation

Signals between the cells of the uterus and the trophoblast cells of the blastocyst encourage the trophoblast cells to physically attach to and begin to invade the uterine epithelium. This is the beginning of the placenta. There are many similarities between trophoblast invasion of the uterine epithelium and a cancerous tumor. A principal difference is that growth stops once the appropriate size of the placenta is reached, whereas a malignant tumor continues to grow. Once implantation has occurred, the embryo nearly doubles in size daily and organogenesis begins, as discussed in Chapter 9.

Blastocyst Development In Vitro

It is important to remember that no existing laboratory conditions support fully normal development of preimplantation embryos of any species. First studied in the mouse, rabbit, and hamster, several characteristics of in vitro–developed embryos have become clear.

FIGURE 8.3

Comparison of mouse blastocysts developed in culture (a) and in the mouse (b).

(a)   (b)

First, blastocysts developed in culture contain markedly fewer cells than blastocysts developed in utero (Figure 8.3). An analysis of the cell cycles of cultured mouse embryos reveals that the zygote stage is approximately the same length, but the two-cell stage is a few hours longer, as is the four-cell stage. The net result of longer cell cycle times during the first few cleavages is that blastocysts form at the species-appropriate time after egg stimulation, but they contain only one-half to one-third the number of cells. For example, hatching blastocysts developed normally in the mouse have approximately 150 cells 120 hours after fertilization. In contrast, mouse blastocysts developed in the laboratory have only one-half to one-third that number of cells. Comparisons of DNA polymerase activities in mouse embryos developing in culture and developing in the oviduct or uterus reveal similar levels per embryo, so a lack of DNA-synthesizing enzymes is not the cause of the increased cell cycle times and the decreased cell numbers in the cultured embryos.

Second, the percentages of cells that make up the ICM remain the same in cultured blastocysts, approximately 30% for the mouse. Because there are fewer total cells, the net result is that cultured blastocysts contain ICMs with fewer cells than embryos developed in utero. Some blastocysts developed in culture contain no ICM cells at all; they are just spheres of trophoblast cells. It is possible that such spheres are analogous to the "empty sacs" observed in some pregnant women whose pregnancies never develop past early implantation. Historically, obstetricians have referred to such early miscarriages as "blighted ovum" in reference to trophoblast cells with no attendant embryonic cells. Given the potential for eggs to activate and initiate cell cleavage under a variety of conditions, there could be multiple reasons for such defective development.

Third, blastocysts that develop in vitro rarely, if ever, develop a layer of endoderm on the inner aspect of the ICM (Figure 8.3). This could simply be because the numbers of ICM cells never reach the critical mass necessary to stimulate the second differentiation event. The fact that primitive endoderm formation is retarded may actually be an advantage for the derivation of lines of ES cells from ICM cells with the potential to differentiate into endodermal structures. Alternatively, it is possible that successful lines of stem cells are only derived from ICMs that do develop primitive endoderm in culture, but not in sufficient cell numbers to be identified as endoderm.

Fourth, it is important to emphasize that a higher percentage of fertilized, or otherwise activated, eggs develops to blastocysts in culture than in the uterus. Several studies have demonstrated that most of the fertilized eggs in a hormone-treated mouse develop to blastocysts in culture, but many fail at the early morula stage in the uterus. It is as if uterine signals stimulate programmed cell death of defective embryos. This is an important area of research that has received little attention but could provide valuable clues about normal developmental signaling between the uterus and preimplantation embryos.

Embryonic Stem Cells

In a landmark report more than 20 years ago, Martin Evans and Martin Kaufman, at the University of Cambridge, England, reported the isolation of pluripotent ES cells from mouse blastocysts. They reasoned that prior attempts to obtain lines of continuously growing nondifferentiated cells had failed because insufficient numbers of ICM cells were obtained and because culture conditions had been more supportive of cell differentiation than of cell division. They took advantage of a fascinating aspect of embryonic development that exists in a number of mammalian species, that of **delayed implantation** (Figure 8.4).

Many mammals, such as rodents and migratory marine mammals, ovulate within a few hours of giving birth, presumably in response to the dramatic drop in pregnancy hormones. Those eggs may be fertilized if she mates with a fertile male, but as long as the mother is producing milk, the fertilized eggs do not implant in the uterus. This is termed **lactational delay** or **embryonic diapause**. The fertilized eggs exist in the uterus as free-floating blastocysts, many of which hatch from the zona pellucida and gradually increase in cell number. Overall, their metabolism slows down. Delayed mouse blastocysts have decreased DNA and RNA polymerase activities, reminiscent of cells that have entered G0. Most of the increase in cell number occurs gradually in the ICM, thus producing blastocysts with especially robust ICM numbers, including some differentiation of primitive endoderm. Under normal circumstances, when the mother weans her pups, the decrease in the hormones responsible for milk production leads to an outpouring of estrogen from the ovary, which in turn stimulates implantation of the delayed blastocyst and the resumption of development. This is an impressive example of the powerful adaptive mechanisms that have evolved to ensure survival of many species.

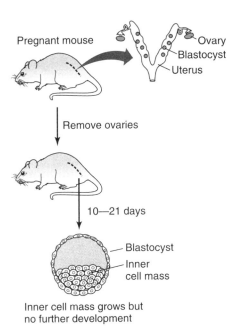

FIGURE 8.4
Schematic of delayed implantation of blastocysts, which occurs naturally in lactating females and can be simulated in the laboratory.

Delayed implantation can be mimicked in laboratory mice by the administration of hormones to a pregnant mouse. The blastocysts exist in a free-floating state for up to 3 weeks under these conditions. Evans and Kaufman recovered delayed implanting blastocysts and placed them in culture. Within days, the blastocysts had attached to the culture dish and the ICM cells had formed large egg cylinder-like structures made up of small round cells surrounded by endodermal cells. The egg cylinders were dispersed into individual cells and the cell suspensions added to culture dishes that already contained a layer of fibroblasts as a feeder layer (discussed in Chapter 6). Actively proliferating colonies of cells were apparent from an early stage, and mass cultures of the ICM cells could be produced.

These investigators previously studied the cell lines produced by culture of ovarian dermoid cysts that occur spontaneously in some strains of mice, as discussed in Chapter 4. Although not all such cell lines were derived from malignant tumors, they are generally referred to as **embryonal carcinoma** (**EC**) cells and are discussed in Chapter 9. EC cells proved to be very useful experimental systems in many respects, but two important features, their lack of uniform chromosome number and their lack of a Y chromosome, limited their usefulness for some types of experiments. It is for this reason that Martin and Kaufman were keen to derive a cell line with a normal chromosome number from a male embryo. One of the first studies they did with the cell lines derived from the delayed implanting blastocysts was to measure chromosome number, which they found to be normal. They termed these new cells **EK cells** to distinguish them from EC cells, but that acronym was quickly replaced by **ES cells**, for **embryonic stem cells**.

The same year, another developmental biologist, Gail Martin, also successfully cultured mouse ICM cells to derive a pluripotent stem cell line. She chose culture medium conditioned by EC cells instead of a monolayer of cultured cells. Dr. Martin had previously studied EC cells and had contributed to the understanding that EC cells were pluripotent by creating offspring that were **chimeric;** this is described in the section to follow.

Importantly, the ease with which ES cells can be derived from the ICM of mouse blastocysts is highly strain specific. The mouse strain chosen by Evans and Kaufmann and Martin, 129J, has proven to be the most successful for deriving ES cells. To date, the reason for this is not known, nor is it known if it is related to the high rates of spontaneous activation of eggs in this mouse strain that lead to the dermoid cysts and teratomas discussed in Chapter 4. Understanding the differences in ease of ES cell derivation between mouse strains may be critical to understanding how to derive therapeutic stem cells through nuclear transfer.

Chimeras

An interesting experimental technique reported independently in 1961 by Andrei Tarkowski, working in Poland, and in 1962 by Beatrice Mintz, working in Philadelphia, involved combining the blastomeres from two cleaving mouse eggs into one mass before compaction (Figure 8.5a). The resulting morula, a **chimera,** developed into a blastocyst with cells from each original egg in both the trophoblast and the ICM. This elegantly simple experimental approach demonstrated not only the pluripotential nature of early cleaving blastomeres, but also established the possibility of obtaining mice with tissues made up of two genetically independent

FIGURE 8.5
Construction of mouse chimeras.

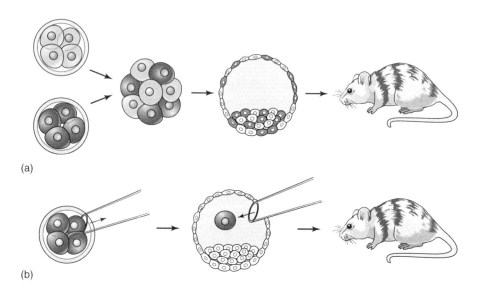

(a)

(b)

The Blastocyst and Inner Cell Mass Cells

cell types. This has proven to be a powerful experimental technique with which to explore the contribution of individual cells to overall development of the mouse.

An early modification of the technique was reported by Richard Gardner in 1968 in which chimeric offspring were obtained after injection of blastomeres into the blastocoel cavity of a mouse embryo just before implantation (Figure 8.5b). Viable offspring were obtained whose tissues were chimeric and included both the cells arising from the fertilized egg and those injected into the blastocoel cavity. This work not only provided further proof of the pluripotency of early embryonic cells but also provided evidence of the accepting nature of ICM cells to the incorporation of "foreign" cells into fetuses. This remarkable plasticity of embryonic development had not been previously appreciated and heralded the new era of manipulating individual mouse genes to more fully understand their function (see sidebar "Knockout Mice").

Genes of Known Importance to Blastocyst Development

A variety of studies revealed genes important and/or essential to blastocyst development. An example of a gene important to the timing of early cleavages and blastocyst appearance is the *Ped* (preimplantation embryo development) gene, discovered by Carol Warner during studies to create chimeras between two strains of mice with differing immune cells. Embryos from one of the strains consistently cleaved more slowly and required more hours for blastocyst formation. The slow trait was eventually linked to the absence of a gene within a region encoding a number of proteins important to the immune system. The function of *Ped* within blastomeres is not known with certainty, but embryos expressing the *Ped* gene product have a selective advantage during development. *Ped* may be just one example of a class of genes that are not essential to reproduction but confer a selective advantage. It is tempting to speculate that *Ped*-positive mouse strains may give rise to more robust ES cells than *Ped*-negative strains.

Examples of genes essential to blastocyst development are listed in Table 8.3. In most cases, the precise cellular function of the gene product is not known with certainty and is an area of active investigation. Some, but not all, of the essential genes are specifically expressed in either the trophoblast or the ICM.

Parthenote and Nuclear Transplant Blastocysts

Animal studies indicate that development to blastocysts by parthenotes and nuclear transplanted eggs is the same as fertilized eggs. Because parthenote and nuclear transplanted blastocysts are developed totally in vitro, it follows that they have delayed cycle times in early cleavages, in keeping with the observed response of fertilized eggs developed in culture. This leads to fewer cells in the ICM, thus challenging the possibility of deriving a line of stem cells. Attempts to derive parthenote stem cells from human parthenotes have reported difficulty in obtaining blastocysts with robust ICMs. No human parthenote stem cell lines have been reported, in contrast to the derivation of parthenote stem cell lines, Cyno-1, from an unfertilized monkey egg.

Knockout Mice

The ability to modify specific genes to determine their function and importance to the organism, as illustrated by the mutant yeast experiments described in Chapter 2, has been widely appreciated by mammalian biologists for decades. One approach to this was the development of a technique termed gene targeting by homologous recombination.

The concept is elegantly simple: Fragments of DNA are taken up by cells in culture and occasionally recombine with their homologous regions in the cell's own DNA. Although rare, such recombination is not as rare as a spontaneous mutation in a single gene. The DNA fragments can be engineered to contain mutations in specific genes as well as DNA sequences that encode products that can be detected to determine the success of the recombination event. For example, a bacterial gene for an enzyme, beta-galactosidase, can be inserted within a mammalian gene, such as the one that encodes the enzyme hypoxanthine-guanine phosphoribosyl transferase (HPRT), which is important in preventing Lesch-Nyan syndrome. This type of DNA construct would result in a gene that cannot express HPRT but can express the bacterial beta-galactosidase that is detectable by a simple staining procedure. Because the ends of the construct would be homologous to the normal HPRT gene, it could replace the normal gene during DNA replication by homologous recombination, hence the term gene-targeted homologous recombination. T cells in culture, including mouse ES cells, actually become genetically modified in this manner, as was demonstrated by several groups of scientists throughout the 1980s.

The next step was to determine whether mouse ES cells, genetically modified by gene-targeted homologous recombination, could contribute to the formation of chimeric offspring. Three groups of scientists reported their results in 1989.

The first report appeared in early 1989 from the laboratory of David Melton in Edinburgh, Scotland [1]. He and his colleagues studied the mouse model for Lesch-Nyan syndrome, animals genetically deficient in the HPRT gene. They derived ES stem cells from the blastocysts of HPRT gene–deficient mice and corrected the gene defect by adding a DNA construct containing a normal gene. They selected those ES cells in culture that had incorporated the new DNA construct into their genome and injected the ES-HPRT-normal cells into the blastocele cavities of embryos that were HPRT deficient. The injected blastocysts were returned to pregnant mothers for gestation and the offspring tested for the presence of the normal HPRT gene. Those offspring that contained the normal gene were then mated and their offspring also tested for the normal HPRT gene. Those animals that could pass the normal gene to offspring obviously had at least some germ cells that contained the normal gene. Such animals are termed founders. This report laid the foundation for correcting genetic defects with ES cell technology.

Two additional reports appeared in the fall of 1989. One was from the laboratory of Bruce Spiegelman, Harvard Medical School, in collaboration with Michael Greenberg, also of Harvard Medical School, and Virginia Papaioannou, Tufts Medical School. This team sought to determine whether genes that were not expressed in mouse ES cells could be targeted by homologous recombination. They chose three genes (*c-fos*, *adipsin*, and *aP2*), whose protein products were not detected in mouse ES cells, for gene targeting by homologous recombination. They mutated the three genes in DNA constructs so as to render them nonfunctional. They then selected ES cells that had undergone homologous recombination with the nonfunctional genes for injection into normal blastocysts. Offspring that contained the mutated genes were then mated with the hope of producing offspring that were negative for the genes. Their experiments were successful [2], and they were able to derive founders with each of the three genes "knocked out" in their germ cells. These experiments demonstrated the power of this approach in studies of function and importance of specific genes.

The third report appeared in November 1989 from the laboratories of Stephen Goff and Elizabeth Robertson, Columbia University, New York [3]. These investigators generated mice with mutations in a gene known to play an important role in many cancers, *c-abl*. Although relatively rare, this team was able to obtain several founder animals with specific mutations in various regions of the *c-abl* gene. Most offspring homozygous for the mutated gene did not survive development, thus illustrating the importance of *c-abl* in normal development. These experiments demonstrated the ability to knock out various regions of genes to study gene function.

These elegant experiments demonstrated the power of knockout mice to help determine the specific roles of specific genes and paved the way for applying the technique of gene-targeted homologous recombination to correct genetic defects in ES cells derived from individuals with genetic diseases, such as diabetes.

REFERENCES

1. Thompson, S., Clarke, A. R., Pow, A. M., Hooper, M. L., and Melton, D. W. (1989). Germ line transmission and expression of a corrected HPRT gene produced by gene targeting in embryonic stem cells. *Cell* 56: 313–321.

2. Johnson, R. S., Sheng, M., Greenberg, M. E., Kolodner, R. D., Papaioannou, V. E., and Spiegelman, B. M. (1989). Targeting of nonexpressed genes in embryonic stem cells via homologous recombination. *Science* 245: 1234–1236.

3. Schwartzberg, P. L., Goff, S. P., and Robertson, E. J. (1989). Germ-line transmission of a c-abl mutation produced by targeted gene disruption in ES cells. *Science* 246: 799–803.

Summary

The first cell commitment event occurs when one blastomere becomes completely enclosed within the morula. The molecular mechanisms controlling this commitment are poorly understood. The ICM cells retain the potential to develop into all the cells and tissues of the developing organism. Developmental pathways are remarkably plastic considering the need for accuracy in differentiation events.

This apparent conundrum may be resolved by the existence of equivalence groups, which are groups of cells with a primary destiny and at least one secondary

TABLE 8.3	Genes Essential to Normal Blastocyst Development
Gene	**Defect if Gene Product Is Lacking**
SRp20	Morula arrest
Ts (tail short)	Morula arrest
Vav	Blastocyst does not hatch
Brg1	Blastocyst does not hatch
PL16	Abnormal blastocyst
CGBP	Abnormal blastocyst
Thioredoxin	Abnormal blastocyst
Ubp2	Abnormal blastocyst
Evx1	Defective trophoblast
Cdx2	Defective trophoblast
b-1 integrin	Defective ICM
Fgf4	Defective ICM
Fgfr2	Defective ICM
B-myb	Defective ICM
Lamc1	Defective ICM
Oct4 (Pou5f1)	Defective ICM
Nanog	Defective ICM

destiny, if not needed for the primary destiny. Thus tissues have more than one opportunity to gain the needed number of committed cells. The existence of equivalence groups may also provide a mechanism for the incorporation of foreign blastomeres into a wide variety of tissues after injection into the blastocoel cavity. This not only proves the uncommitted status of blastomeres but also the lack of rejection of foreign embryonic cells introduced during development. It is this very plasticity that predicts the therapeutic value of ES cells.

Stem cells with the capacity to continue to renew in laboratory culture are derived from the ICM of blastocysts. In theory, blastocyst development should be comparable from eggs activated by fertilization, parthenogenesis, or after nuclear transplantation. Nonetheless, as of this writing, neither human parthenote blastocysts nor human nuclear transplant blastocysts have yielded stem cells.

Technological advances in detecting gene expression in small numbers of cells have provided valuable new information about cell functions essential to each developmental milestone. It is possible that similar pathways may be important in directing the fate of pluripotent stem cells.

Review Questions (True or False)

1. The first commitment event in mammalian development occurs when inside cells in the morula become entirely surrounded by outside cells.
2. Some developmental events are independent of the number of cell cycles after egg activation.
3. The same genes are expressed in the trophoblast cells as in the ICM cells.
4. Pluripotent stem cells are derived from ICM cells that become adapted to laboratory culture conditions.
5. Chimeras develop from combining blastomeres from two or more embryos into one embryo.
6. Trophoblast cells and ICM cells express the same genes.
7. Even though activated eggs may progress to the blastocyst stage, they may not contain an ICM.

ADDITIONAL READINGS

Adjaye, J., Huntriss, J., Herwig, R., BenKahla, A., Brink, T. C., Wierling, C., Hultschig, C., Groth, D., Yaspo, M.-L., Picton, H. M., Gosden, R. G., and Lehrach, H. (2005). Primary differentiation in the human blastocyst: comparative molecular portraits of inner cell mass and trophectoderm cells. *Stem Cells* 23: 1514–1525.

Chung, Y., Klimanskaya, I., Becker, S., Marh, J., Lu, S.-J., Johnson, J., Meisner, L., and Lanza, R. (2006). Embryonic and extraembryonic stem cell lines derived from single mouse blastomeres. *Nature* 439: 216–220.

Comiskey, M., Goldstein, C. Y., DeFazio, S. R., Mammolenti, M., Newmark, J. A., and Warner, C. M. (2003). Evidence that HLA-G is the functional homolog of Qa-2, the Ped gene product. *Hum Immunol* 64: 999–1004.

Edwards, R. G. (2000). Human embryos in vitro: pioneer illustrations of oocyte maturation, fertilization, cleavage and blastulation. *Hum Reprod* 15(Suppl 4): 1–9.

Hartshorne, G. (2000). The embryo. *Hum Reprod* 15(Suppl 4): 31–41.

Johnson, M. H. (1977). *Development in Mammals*. Amsterdam: Elsevier/North Holland.

Kiessling, A. A., Davis, H. W., Williams, C. S., Saute, R. W., and Harrison, L. W. (1991). Development and DNA polymerase activities in cultured pre-implantation mouse embryos: comparison with embryos developed in vivo. *J Exp Zool* 258: 34–47.

Warner, C. M., Brownell, M. S., and Rothschild, M. F. (1991). Analysis of litter size and weight in mice differing in Ped gene phenotype and the Q region of the H-2 complex. *J Reprod Immunol* 19: 303–313.

Wolstenholme, G. E. W. (1965). *Preimplantation Stages of Pregnancy*. Boston: Little, Brown.

CHAPTER 9

Organogenesis

Mammalian parthenotes cannot be induced to develop normally to term.

Tomohiro Kono, 2002

OVERVIEW

It seems so unlikely that a fertilized egg could give rise to an entirely new individual complete with functioning organs and brain that the mere notion is preposterous, and yet it happens over and over. Aristotle attempted to understand embryogenesis through the study of fertilized chicken eggs. The developing chicken, like amphibia, has provided some fundamentally important insights into organogenesis because many characteristics of the early stages of embryogenesis are highly conserved across species. Mammals, however, form placentas and thus require their own detailed study.

John Hertig was a pathologist colleague of Ruth Menkin and John Rock (see sidebar, Chapter 4). To more fully understand early human development, Hertig systematically sectioned the uteri of women undergoing hysterectomy by John Rock. Dr. Rock explained his interest in early human development to dozens of women in need of hysterectomy for unrelated reasons. The women agreed to have a hysterectomy at timed intervals after ovulation and intercourse. Dr. Hertig then explored the entire uterus for the presence of an early pregnancy, which was then examined. In this way, Hertig and Rock compiled the most complete description of early human development available to this day. Most of their original materials are in storage at the Carnegie Institute. Given the current social concerns with studies of early fertilized human eggs, we are deeply indebted to their work, for without it we would understand even less than we do now.

In 1956, Hertig, Rock, and Adams described a unique series of "34 human ova within the first 17 days of development." Hertig summarized his many years of examining early developmental stages as follows: 15% of eggs ovulated by normal fertile women fail to fertilize and another 10 to 15% fertilize but fail to implant. Of the 70 to 75% of fertilized eggs that implant, only 58% survive until the second week and 16% of those are abnormal. Thus, only 42% of eggs exposed to sperm survive until the first missed menstrual period, and of those a number are aborted during subsequent weeks. These findings are in basic agreement with more recent large epidemiologic studies that indicate that it takes an average of 3 to 4 months for normal fertile couples to achieve a viable pregnancy.

It is important to note that at no time were the fertilized eggs referred to as "embryos" by Hertig or Rock. As discussed in Chapter 4, the term *embryo* was not applied to cleaving eggs until after assisted reproduction became the standard of care for infertile couples. Interestingly, the term *zygote* was broadly applied to early developmental stages, sometimes even after implantation. The term *zygotic genome activation* (ZGA) has been used to refer to the turn-on of gene expression in cleaving blastomeres. Implied in this use of the term is the notion that an embryo is a stage reached after developmental control is transferred entirely from the egg cytoplasm stores to the new genome created as a result of fertilization. The term is also applied to describe the two types of twins that may arise: fraternal twins (**dizygotic**) arising from two fertilized eggs and identical twins (monozygotic) arising from one fertilized egg that may split into two developing embryos at various stages after the first cleavage and before the formation of the primitive streak, as described below.

Dizygotic: twins arising from two fertilized eggs.

The Second Week

As described in Chapter 8, implantation occurs approximately 7 to 10 days after fertilization. The trophoblast cells overlying the inner cell mass (ICM) attach to the newly enriched layer of cells that line the uterus (endometrium) and begin to invade between them (Figure 9.1). The trophoblast secretes human chorionic gonadotropin (hCG), a protein hormone that maintains the corpus luteum, as described in Chapter 3. Human chorionic gonadotropin may also be a growth factor that stimulates development of the placenta. Once the placenta has reached a critical mass of cells, it synthesizes and secretes the steroid hormones, estrogen and progesterone, necessary to maintain the pregnancy.

A layer of primitive endoderm forms on the inner aspect of the ICM, the remaining cells of which are now sometimes termed *primitive ectoderm*. By the end of the second week, another cavity begins to appear between the primitive ectoderm and the overlying trophoblast. The net result is that the ICM takes on the appearance of a pancake, two cell layers thick, attached at the periphery to trophoblast with a small cavity above and the blastocoel cavity (primitive yolk sac) below (Figure 9.1).

FIGURE 9.1
Early implantation and formation
of the bilaminar germ disc.

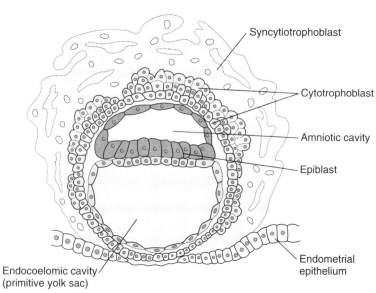

This stage is the bilaminar germ disc or epiblast, which contains on the order of a few hundred embryonic stem cells. Very little is known about gene expression and controls on development at this stage in humans.

The trophoblastic cells develop rapidly during this stage and differentiate several types of cells that form the complex and protective cell layers of the placenta, including the chorionic villous membrane, which interacts with maternal blood circulation to take in nutrients and release fetal metabolic byproducts. As discussed below, mature placentas may contain a substantial population of embryonic stem cells.

The Third Week

At approximately the time of the first missed menstrual period, the bilaminar germ disc becomes a trilaminar germ disc. **Mesoderm** appears as a population of small rapidly dividing cells between the primitive endoderm and the embryonic ectoderm. Mesodermal cells are protected from the outside environment. They initially create two bulging regions that have a slight depression between them, termed the **primitive streak**, that extends from one edge of the disc to approximately the middle, where it stops at a depression termed the **primitive node**. The appearance of the primitive streak defines the orientation of the developing embryo.

The embryonic ectoderm is the **dorsal** (back) side. The primitive node is the lowest point of the spinal cord. Directly opposite the primitive streak is the region that eventually gives rise to the head. Hence, after approximately 15 days of development, the embryo has a right and left side as well as head (**cephalic**) and tail (**caudal**) regions and a top (dorsal) and bottom (**ventral**). Things begin to happen even faster. The mesoderm, which in many ways is reminiscent of the original ICM because it has little contact with the outside environment, expands rapidly, causing a thickening and reorganization of the disc. From the primitive node to the head region, a tube-like area forms within the mesoderm, the **notochord**, which is the precursor to the spine. The embryonic disc takes on a pear-shaped appearance, with most of the expansion in the cephalic region; the primitive pit is thus displaced to the caudal third of the disc instead of in the middle (Figure 9.2).

Continued expansion of the mesodermal layer creates a new midline groove in the cephalic region. The ectodermal cells outgrow on either side of the midline groove to define the **neural plate**, at the apex of which are the **neural crest cells**. This dorsal-reaching growth phase creates ectodermal flaps that reach up and toward each other, eventually touching at the top and fusing at the midline in the center of the embryo. The net result of this growth is to capture a tube of ectodermal cells (the **neural tube**) inside the mesoderm, with a population of neural crest cells located just under the overlying ectoderm (Figure 9.3).

At the same time the neural tube is forming, bulges of mesoderm are creating **somites**, which appear like two rows of corn kernels on either side of the neural tube. The somites are mesodermal cells that expand in clusters, forming organizing regions

Primitive streak: the depression in the ectodermal layer of the embryonic disc caused by mesodermal cells.

FIGURE 9.2
The third week.

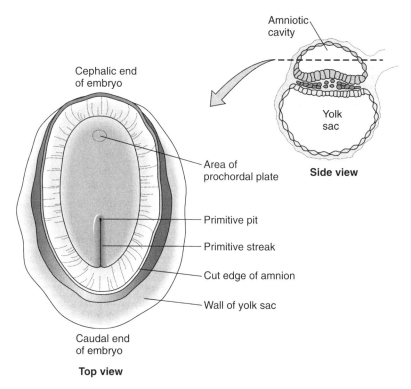

Amniotic cavity

Cephalic end of embryo

Area of prochordal plate

Yolk sac

Side view

Primitive pit

Primitive streak

Cut edge of amnion

Wall of yolk sac

Caudal end of embryo

Top view

for the embryo, which will eventually give rise to skeletal structures such as vertebra and ribs.

The regularity and symmetry of somite appearance are an example of the precision necessary for successful fetal development. The molecular guides for such highly orchestrated processes are largely unknown. Information from fly and

FIGURE 9.3
Formation of the neural tube.

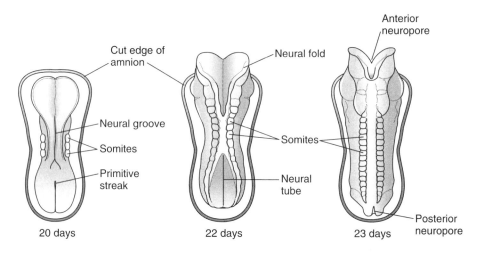

Cut edge of amnion

Neural fold

Anterior neuropore

Neural groove

Somites

Primitive streak

Somites

Neural tube

Posterior neuropore

20 days

22 days

23 days

chick embryos suggests somite formation involves gene expression with a precise periodicity. Genes that induce division of mesodermal cells to form the somites are expressed in a pulsatile fashion for a defined period of time (e.g., 30 minutes for fly embryos) in the location of the growing somite. A subsequent 30-minute interval grows the next somite pair and so on. The periodicity is substantially longer for mammals. Understanding the molecular basis for such periodicity may reveal what guides symmetry and growth throughout development. The same pathways may be essential to controlling differentiation of cultured pluripotent stem cells in laboratory conditions.

The forward growth of the developing embryo has greatly elongated it and positioned the primitive node within the caudal one-fifth. As the somites appear, the neural tube closes forward and aft, leaving a large and growing opening at the cephalic end, with a much smaller opening at the caudal end. The ventral aspect of the embryo is attached at the periphery to the trophoblastic cells (now the amnion) and in the center to what is left of the blastocoel, which has now become the yolk sac. Just under the neural tube at the cephalic end, the mesodermal cells have formed a bulge that will give rise to the heart.

Most of the growth of the pregnancy to this stage is in the placenta and extraembryonic membranes. The developing embryo has no heartbeat and only 10 or 12 somites, approximately one-third of the total number that ultimately form to organize the skeletal structures. The bulk of the embryonic cells are stem cells that have undergone few if any differentiation events, although their commitment to ectoderm, mesoderm, and endoderm may be fixed by this stage. The neural crest cells have been studied in several species because they ultimately give rise to many different cell types. Although they are ectodermal in origin, they migrate in unique ways within the mesoderm. There is something about being enclosed entirely within the embryo, not exposed to the outside, that serves to maintain the pluripotency of embryonic cells.

This may relate to the concept of equivalence groups, described in Chapter 8. Equivalence groups are a population of cells poised to establish their next developmental fate. They may, as a group, have undergone an initial commitment decision (e.g., to become mesoderm instead of ectoderm). A subset of the cells is recruited to accomplish the next developmental milestone, such as heart muscle cells. If heart muscle cells successfully develop, the remaining cells in the equivalence group are available for the next developmental need. If heart muscle cell development is not successful, additional members of the equivalence groups are recruited to try again.

This paradigm predicts the existence of cellular factors that maintain the multipotency of the equivalence groups, such as those within newly formed somites. Current evidence suggests this is accomplished by cell-to-cell contact. The intracellular pathways activated by such contact maintain expression of the genes responsible for the lack of differentiation of the cells in the equivalence groups. Such pathways may also be important to the maintenance of pluripotency in cultured stem cells.

The Fourth Week

It is interesting to note that most medical embryology textbooks refer to the fourth to eighth weeks of development as the embryonic period. As discussed previously, the first 3 weeks are sometimes referred to as the zygotic (as in "fertilized egg") period. This terminology serves to emphasize that in the minds of those who study development, the embryo does not exist until at least 21 days after fertilization.

Embryonic stem cells divide approximately once a day, so the developing embryo almost doubles in size daily. This becomes especially obvious during the fourth week. The nutritional requirements of the developing tissues exceed what can be obtained from the cavities surrounding the embryo, so further development depends on successful formation of the primitive heart and the gut (Figure 9.4).

The precursor cells to the blood system come together in the cardiac bulge and form the first heart tube, which begins to pulsate toward the end of the fourth week. It will be several weeks more before the four-chambered heart is fully formed, but the early heart structures exhibit rhythmic contractions that serve to force embryonic blood cells through developing capillaries.

The ventral aspect of the embryo has developed a primitive gut by bringing the lining of the **yolk sac** (formerly the blastocoel) into closer approximation to the ventral endoderm of the embryo. As early as the end of the fourth week, liver and lung buds are distinguishable as outgrowths of the endoderm. The connection to the yolk sac becomes increasingly restricted as a stalk is formed adjacent to the connecting stalk that will become the umbilical cord.

Most of the somites appear by the end of the fourth week. The rapid growth of the neural tube, the primitive central nervous system, has almost closed the cephalic end of the neural tube to form the beginning of the primitive brain cavities. Formation of the central nervous system occurs gradually, and brain development continues well into the fetal period, extending nearly to the time of delivery. Many embryonic and fetal systems function well in advance of even primitive brain development.

FIGURE 9.4
The fourth week.

Anterior neuropore

Cut edge of amnion

Connection to yolk sac

Posterior neuropore

25 days

Optic placode

Lens placode

Heart bulge

Vitelline duct

Allantois

Umbilical cord

Limb buds

28 days

Fifth and Sixth Weeks

The rapid growth of the developing embryo incorporates the endodermal structures completely within the body cavity (Figure 9.5). Lung, liver, stomach, pancreas, and bladder are distinguishable. The growth factors and signals that stimulate their development are largely unknown. The primitive kidney arises in the mesoderm and is also distinguishable by the sixth week.

By the end of the sixth week, the primordial germ cells have migrated by ameboid movement from what is left of the yolk sac to the gonadal ridge, where their appearance stimulates the development of the testis or the ovary. Expression of Y chromosome genes induces development of the testis and epididymis. In the absence of a Y chromosome, an ovary, oviduct, and uterus develop.

Five weeks

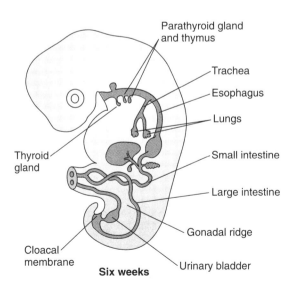

Six weeks

FIGURE 9.5
The fifth and sixth weeks.

The developing embryo is about half an inch long and weighs approximately half an ounce. Tissue and organ growth is proceeding exponentially, principally because of the expansion of stem cell populations. Some stem cells have become committed to specific tissue types, whereas others remain pluripotent. Few cells are fully differentiated. That process requires most of the 38 weeks of fetal development.

Primordial Germ Cell Migration

Although the development of all organ systems is nothing short of miraculous, preservation of family genetics depends on the successful organization of fetal ovaries and testes. The link between ancestors and descendants depends on the small group of cells, called **primordial germ cells (PGCs)**, that make the journey from the yolk sac into the mesodermal region, termed the **gonadal ridge**, at the end of the sixth week (Figure 9.6). The small band of participants and the name of their destination is remarkably reminiscent of the pioneers that settled the American West in the 1800s.

PGCs: primordial germ cells, the progenitors of sperm or eggs.

First described in mice (see sidebar), PGCs express higher than average levels of an easily detected enzyme activity, alkaline phosphatase, an enzyme that removes phosphate groups (see Chapter 2) at pHs higher than neutral (see Chapter 5). The role for alkaline phosphatase in PGC physiology is not known, but it has provided a useful PGC marker for numerous tissue studies. In the mouse, approximately 100 PGCs are estimated to reside in the mesoderm on

FIGURE 9.6
Migration of primordial germ cells.

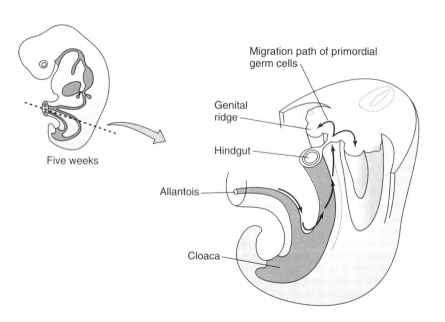

Five weeks

Migration path of primordial germ cells

Genital ridge

Hindgut

Allantois

Cloaca

Primordial Germ Cells and Embryonal Carcinoma Cells

Nearly 50 years ago it was discovered that the strain of mouse that had a relatively high incidence of ovarian dermoid cysts also exhibited spontaneous testicular teratomas in 1% of males. This discovery was made by L. C. Stevens and C. C. Little at the Jackson Laboratories. During the 1960s, this team reported several characteristics of the teratomas that laid the groundwork for current understanding about both embryonic carcinoma cells and embryonic germ cell stem cells.

First, the incidence of teratomas was markedly increased by transplantation of the gonadal ridge from fetuses into the testes of adult mice. Importantly, the incidence of teratomas from other strains of mice could also be increased by transplantation of the gonadal ridge into adult testes. However, tumor induction was limited to a 2-day interval during development, when many of the germ cells had completed the migration. Once the migration was completed (approximately days 12–13 of gestation), the incidence of tumor formation decreased dramatically. Thus, the ability of germ cells to continue cell division was markedly reduced once the gonadal ridge was fully populated, implying some change in the pluripotency of the cells.

Second, some teratomas could be retransplanted, but others could not. The retransplantable tumors, termed *teratocarcinomas,* contained embryonic carcinoma cells with morphology characteristic of human teratocarcinomas.

Third, both transplantable and nontransplantable tumors contained differentiated tissue elements, such as bone, cartilage, hair, and teeth, similar to elements found in ovarian tumors. Importantly, however, work by L. J. Kleinsmith and G. B. Pierce in 1964 demonstrated that a single cell from a transplantable tumor was found to be sufficient for the growth of a new tumor with the full range of differentiated tissue elements noted in the original tumor.

Taken together, these experiments established that embryonic germ cells could give rise to pluripotential tumors, some of which could become malignant cells with the potential to give rise to multiple cell types. In many respects, these experiments were the first demonstration of embryonic stem cells.

During the 1970s, Davor Solter and his colleagues reported a series of experiments demonstrating that teratocarcinomas also resulted from transplantation of mouse blastocysts to testes or kidneys. Established cell cultures from mouse embryonic carcinoma cells were first reported in 1970 by B. W. Kahn and B. Ephrussi. That report was quickly followed by confirmatory work from several laboratories, establishing the pluripotent nature of cells derived both from germ cells and from preimplantation mouse embryos.

By the end of the 1970s, it had been established that both primitive germ cells and mouse blastocysts could give rise to tumors with pluripotential cells that could be induced to differentiate into a variety of cell types. In fact, it was difficult to prevent cell differentiation, just as it is today. Although this work was seminal to present-day stem cell technology, because the origin of most of the cell lines was tumors, the cells themselves were heteroploid, thus limiting their usefulness as stem cells.

day 7, approximately 2 days after implantation. Their migration begins shortly thereafter (analogous to approximately 5 weeks after implantation in the human). Remarkably, they proliferate as they migrate, with estimates of tens of thousands eventually populating the gonadal ridge. They migrate over the course of the following week, which represents two-thirds of the gestation time of a mouse.

The presence of a Y chromosome initiates formation of the testes and the degeneration of early structures that would otherwise give rise to female reproductive tract organs. In the absence of a Y chromosome, ovaries and associated female structures begin to form. Therefore, by the end of the sixth week, most tissues and organs can be identified and are populated by rapidly expanding stem cell populations, dedicated to the formation of a fully functioning organism. The potential number of misdirections possible during this process seems infinite, but the low number of developmental anomalies present in newborns suggests an equal number of corrective balances must be operational throughout fetal development.

Fetal Tissue Stem Cells

Most of the studies done on harvesting embryonic stem cells from fetal tissues have been conducted in mice. The basic premise has been that embryonic stem cell lines derived from specific tissues are restricted to the development of those tissues, for example, heart, spinal cord, and liver. Experiments have been designed, therefore, to harvest stem cells from tissues of interest and develop the laboratory methods for expanding those cell lines in the laboratory. Because one of the hallmarks of stem cells is that they are precursor cells and, as such, do not perform adult cell functions, another aspect of fetal stem cell research has been to understand the factors that bring about their differentiation and normal adult cell function.

For example, nerve stem cells can be isolated from the neural tube described previously and expanded in culture in the laboratory. Their continual division requires the presence of added growth factors, supporting the concept that they do not synthesize the necessary growth factors themselves. Fibroblast growth factor (FGF) seems to be essential for the continued cell division of a wide variety of neural stem cells.

As of this writing, stem cells have been isolated from the developing **forebrain,** the **midbrain,** and the **hindbrain,** and the spinal cord and the developing *cortex* of mice and rats. For the most part, multipotent stem cells do not migrate but appear to maintain a relatively stable population localized to their zones. As some of them begin the process of differentiation, they become more mobile and begin to migrate to selected target sites. This is an example of the many stages of differentiation that some cell types pass through during the course of tissue organization. Each stage may also continue to divide; thus they exist in a multipotent state, more differentiated than their stem cell precursors but more capable of expanding their populations than the mature fully functioning cells they are destined to become.

In addition, stem cells have been identified from the developing retina. Characterization of retinal stem cells has led to the further discovery that adult retinas also contain a small population of self-renewing stem cells, a fact not widely appreciated before fetal retinal stem cells were characterized. This is an example of the powerful discoveries that may be made about adult tissues through studies of embryonic stem cells. It is conceivable that by understanding fetal tissue stem cells, it may be possible to devise treatment strategies that would promote proliferation and repopulation of damaged adult tissues through regeneration of their own stem cell reserves.

Another neural stem cell of particular interest is the neural crest cell, which appears at the margin of the developing neural tube, as described previously. The neural crest cell is of particular interest for a number of reasons. It is capable of differentiating into several different types of cells, including crossing over from an epithelial cell into mesenchymal cells, which are usually derived from mesoderm cell types. Neural crest cells are highly migratory and have been shown to give rise to pigment-producing cells (melanocytes), neurons, glial cells, bone, cartilage, smooth muscle, and blood cells. Studies of this remarkable plasticity have intensified within the past few years in an effort to understand, and take advantage of, the pluripotent nature of this remarkable cell.

In contrast to neural cell precursors that have been isolated from fetal rat and mouse tissues by a number of investigators, stem cells from the earliest discernible islands of blood cell formation (**hemangioblast**) have not been successfully isolated and propagated in culture. This is surprising in light of the fact they were first identified 70 years ago. **Hematopoietic stem cells** have been isolated, however, by a number of investigators from both fetal and adult animals. Lymphocytes, monocytes, neutrophils, red blood cells, and megakaryocytes (give rise to platelets) all derive from the hematopoietic stem cell (see Figure 1.1). In no other system is the powerful multipotential nature of stem cells better illustrated.

Organs deriving from endodermal embryonic cells—liver, pancreas, and gut—contain a population of stem cells in adult tissues. Hence, although studies of fetal stem cells from these organs have been reported, the bulk of the work has focused on the isolation of stem cells from adult organs.

Trophoblast is the first cell to differentiate during early embryogenesis, and it subdivides itself into two distinct cell types very soon thereafter. One type remains associated with the ICM (the polar trophectoderm) and continues to divide. Opposite the ICM, at the abembryonic pole, the trophoblast cells cease to undergo cell division but continue to replicate their DNA, becoming **polyploid**, as described in Chapter 6. These distinctions appear to carry over into mature placenta as well. Given the embryonic nature of trophoblast cells, several groups of investigators attempted to isolate stem cells from trophoblastic tissues, particularly from early mouse embryos. Finally, in 1998, a specific combination of growth factors and feeder layer cells supported the isolation of several lines of

Polyploid: containing more than the normal number of chromosomes.

trophoblast stem (TS) cells. If either the growth factors or the feeder cells are removed, the trophoblast stem cells differentiate into giant cells. In some reports, the feeder layer can be replaced by medium conditioned by the feeder layer (as described in Chapter 6), but these reports are just beginning to appear. The ease with which trophoblast stem cell lines can be established suggests that this is a valuable system for studies to understand the factors needed to maintain the undifferentiated state of stem cells.

Parthenogenesis

There are many examples of parthenogenesis, the development of offspring solely from eggs, throughout the animal kingdom. Aphids are a common example. But mammalian eggs cannot accomplish development without sperm. Some activated mammalian eggs can initiate development, including implantation, but organogenesis fails.

Studies of parthenogenesis in mice have provided an excellent model system with which to begin to understand the role of paternal genes in embryonic organization. Such studies have revealed the importance of gene imprinting in regulating the precise patterns of gene expression necessary for normal embryogenesis. As described in Chapters 4 and 5, imprinting is a modification to a gene that does not change its basic coding sequencing but influences what cells can express the gene. A gene contributed to an embryo by the egg is expressed under different cellular circumstances than a gene contributed by the sperm. The phenomenon of imprinting has been appreciated and documented for a variety of mammals for a couple of decades, but the precise molecular mechanisms remain unclear.

Tomohiro Kono and his colleagues conducted an elegant series of studies of development of mouse parthenotes that contained one haploid set of genes fully imprinted by a mature egg and a second set of haploid genes that were not fully imprinted. They discovered an important role for two genes in the developmental arrest of parthenotes. One gene, termed *H19*, was expressed at too high a level and the other gene, termed *Igf2*, was expressed at too low a level. *Igf2*, **insulin-like growth factor 2** (also known as somatomedin A), was the first imprinted gene identified in mammals and has been widely studied by a number of investigators. It is located on chromosome 11 within a cluster of genes that also includes *H19*. *Igf2* is known to be expressed from the paternal genome in developing mouse embryos and is an important growth factor. *H19* is an RNA that is not translated into protein but appears to down-regulate *Igf2* expression through its RNA sequence.

Importantly, Kono and colleagues found that inhibiting *H19* enhanced *Igf2* and resulted in extending development of mouse parthenotes through early stages of organogenesis before their arrest. This exemplifies the critically important balance between maternal and paternal genes necessary for normal development of offspring.

Summary

Studies in the 1950s revealed fewer than 40% of normally fertilized human eggs have the capacity to initiate fetal development. Some may develop to blastocysts and undergo implantation near the end of the first week but fail thereafter.

The survivors develop three layers of embryonic cells, endoderm, mesoderm, and ectoderm, and by the end of 4 weeks a primitive beating heart has developed, marking the transition from the embryonic stage to the fetal stage. By the end of the sixth week, most organs are discernible and germ cell migration occurs, populating the gonadal ridge with eggs or sperm if there is a Y chromosome.

The process of organ formation and cell differentiation persists throughout the 48 weeks of fetal development. Successful development of tissues and organs requires extraordinary coordination of growth signals, stem cell responses, and appropriate cellular migration. Rich in stem cell populations, embryonic organs and tissues should be a reservoir of pluripotential stem cells, few of which have actually been isolated, cultured, and studied, principally due to a lack of knowledge about the laboratory conditions needed to sustain their growth without cell differentiation. Stem cell science may depend on the sequence of gene expression during fetal development to bring about stable differentiation of desired cell types, such as heart muscle cells, in the laboratory to be used for therapeutic purposes.

Review Questions (True or False)

1. Most eggs exposed to sperm survive and properly implant.
2. The mesoderm of the blastocyst does not appear until it is 3 weeks old.
3. Cells in the interior of the embryo maintain pluripotency longer.
4. Corrective mechanisms exist to recruit uncommitted embryonic cells to replace those that fail to properly form various tissues.
5. The brain is one of the first organs to fully form in the developing embryo.
6. Mammalian parthenotes can give rise to fully functioning embryos without fertilization by sperm.
7. Teratomas (tumors containing hair, teeth, bone and cartilage) provided the first demonstration of the pluripotent nature of embryonic stem cells.

ADDITIONAL READINGS

Artavantis-Tsakonas, S., Rand, M. D., and Lake, R. J. (1999). Notch signaling: cell fate control and signal integration in development. *Science* 284: 770–776.

Kono, T., Sotomaru, Y., Katsuzawa, Y., and Dandolo, L. (2002). Mouse parthenogenetic embryos with monoallelic H19 expression can develop to day 17.5 of gestation. *Dev Biol* 243: 294–300.

Langman, J., and Sadler, T. W. (1988). *Medical Embryology*. Baltimore: Williams & Wilkins.

Mammalian Nuclear Transplant Technology

Differential activity of maternal and paternal genomes, and the results presented here, suggest that the cloning of mammals by simple nuclear transfer is biologically impossible.

David McGrath and Davor Solter, 1984

OVERVIEW

The 1984 report by McGrath and Solter was compelling but biologically illogical. Several lines of evidence provided strong support for the concept that embryonic and adult development did not fundamentally change the chromosomal makeup of individual cells in different tissues. Theoretically, all genes were still present in all cells. Why, then, could they not be reprogrammed to direct development of a new individual?

Jim Robl at the University of Wisconsin laboratory of Neal First continued to challenge the McGrath and Solter dogma. Robl substituted electrofusion for the Sendai virus–induced fusion practiced by McGrath. The principle behind electrofusion is similar to Sendai virus–mediated fusion except the plasma membranes are caused to fuse together by a small electrical jolt (Figure 10.1). Robl fused pronuclei with enucleated cow zygotes to derive reconstructed embryos. Two reconstructed embryos were transferred to pregnant recipients and two calves were born, thus demonstrating that the process of manipulation and

Fertilized cow egg

Pronuclei
removed

New
pronuclei
injected

Electricity

Electricity

Electricity

Activation
and cleavage

Activation
and cleavage

Activation
and cleavage

Arrest

Arrest

Calves

No offspring

No offspring

FIGURE 10.1
Transplantation of cow nuclei by
electrofusion.

electrofusion did not significantly alter developmental potential. However, nuclei from two-, four-, or eight-cell embryos fused to enucleated zygotes did not produce embryos that could develop further. This work was published in 1987 and supported the work with mouse zygotes that revealed that the ability of transplanted nuclei to direct embryonic development of is restricted relatively early in mammalian embryos.

A slightly different approach was taken by Yukio Tsunoda working in the laboratory of Professor Sugie. Tsunoda and his colleagues transferred nuclei from four-cell and eight-cell stage blastomeres into enucleated two-cell stage blastomeres

FIGURE 10.2
Transplantation of nuclei from
four-cell and eight-cell stage
blastomeres into enucleated
two-cell stage blastomeres.

instead of zygotes and obtained not only improved development in laboratory culture but also the birth of live offspring (Figure 10.2). This suggested that zygotes
were incompatible with transferred nuclei, in agreement with more recent reports
that as early as 2 hours after fertilization, zygotes cannot remodel sperm heads and
supported the concept that the nuclei from eight-cell stage blastomeres could still
direct full development of new offspring. The problem seemed to lie in incompatibility between transplanted nuclei and zygote cytoplasm.

By the late 1980s, the power of the cell cycle regulators described in Chapter 2
over nuclear events had become widely appreciated. Developmental biologists began
to explore the relationship between the stages of the cell cycle of the zygote and the
nucleus being transplanted.

Nuclear Transplantation and the Cell Cycle of Fertilized Eggs

One of the earliest studies that focused on the role of the cell cycle in successful
blastomere cleavage after nuclear transplantation was conducted by Smith, Wilmut,
and Hunter at the Edinburgh Research Station in Scotland. Mouse embryos were
their experimental system. They reported in 1988 that cross-transfer of pronuclei

from one zygote to another resulted in a higher rate of blastocyst development (94%) if the donor and recipient zygotes were synchronized with respect to fertilization. In the same study, they observed that the timing of the transplantation of two-cell nuclei to enucleated zygotes was more complicated. Both the **karyoplast** (the two-cell nucleus) and the **cytoplast** (the enucleated zygote) influenced cell cleavage and successful development to blastocysts. Overall, reconstructed embryos from two-cell nuclei at later stages of the two-cell cycle (during the long G2 phase described in Chapter 6) transplanted into late stage (G2) zygotes had the greatest potential to develop to blastocysts. This observation suggests that nuclei that had completed S phase and passed through the checkpoint for DNA replication were better prepared to resume division in a strange cytoplasm, especially if the cytoplasm were also at the G2 stage of the cell cycle.

A follow-up study published in 1990 by Smith, Wilmut, and West examined the outcome of fusing early and late karyoplasts from zygotes and two-cell nuclei to early and late zygote cytoplasts. Early pronuclei and two-cell nuclei (probably in early S phase, because we now know that the G1 phase of both pronuclei and two-cell nuclei is short) fused with late zygote or two-cell cytoplasts delayed cleavage for a few hours. This agrees with the notion that nuclei in S phase can complete DNA synthesis in the presence of G2 cytoplasm because the nucleus has gathered within itself all the necessary enzymes and supporting elements for DNA replication. Moreover, G2 cytoplasm has little influence over the nucleus until S phase is completed. In contrast, late pronuclei and two-cell nuclei (in G2) fused with early enucleated zygote and two-cell cytoplasts (probably in the G1/S transition or S phase) resulted in embryos that did not cleave earlier than expected, according to the age of the cytoplasm. This important observation is consistent with the known cyclin/Cdc control of the G2 to M phase transition discussed in Chapter 2. Many late two-cell karyoplasts transferred into early zygote cytoplasts failed to undergo cleavage at all, suggesting that the long G2 phase of the mouse two-cell stage blastomere is required for a specific interaction between the nucleus and the cytoplasm in this species. Whether or not this is true for other species at some developmental stage is not known with certainty.

Cytoplast: the egg cell remaining after the removal of all chromosomes.

Transplantation of Embryonic Nuclei Into Unfertilized Eggs

Of particular importance to the development of human pluripotent stem cells from nuclear transplant into eggs has been the work done in cloning animals. This approach necessitated an improved understanding of egg physiology, including egg activation without sperm. Termed **parthenogenesis** (discussed in Chapters 4, 8, and 9), egg activation requires both overcoming the cellular controls responsible for maintaining arrest at metaphase II (see sidebar, Chapter 4) and then deploying the egg stores responsible for several rounds of chromosome duplication and blastomere cleavage. In addition, because the goal of nuclear transplantation experiments was to test the potential of differentiated nuclei to support embryonic development, the transplanted nucleus could need to undergo remodeling to a dedifferentiated state before cleavage was initiated—a tall order.

Although some reproductive biologists pursued studies of nuclear transplantation into eggs to more fully understand cell differentiation, others were interested in developing methods for cloning animals, especially valuable farm stock, or important disease research models. In this regard, multiple use of the term **clone** has created some confusion. As noted by early developmental biologists (see sidebar, Chapter 7), separation of early cleaving blastomeres could result in the development of multiple offspring. Twins created by separation of two-cell embryos, for example, are relatively common in many species. Steen Willadsen reported a number of studies throughout the 1980s aimed at producing identical twins and triplets from cow and sheep embryos that had been manually bisected or quadrisected (Figure 10.3). Some experiments were successful, and twins and quadruplets developed from single embryos. Such offspring are, strictly speaking, genetic clones of each other.

FIGURE 10.3
Multiple offspring ("clones") produced from single embryos.

2-cell stage

Twin sheep

8-cell stage

Quadruplets

The other use of the term *clone* refers to individuals who result from the transfer of a nucleus from an existing individual (including adults) into an egg that is then induced to initiate embryonic development. Willadsen also conducted early studies of blastomere fusion with unfertilized eggs using sheep eggs and embryos. He reported his findings in 1986. His basic procedure was to bisect unfertilized eggs to remove their genetic information and then fuse the half of the egg with no chromosomes to whole blastomeres from sheep morulae at the 8- to 16-cell stage. Although the success rates were low, a few lambs were born from the reconstructed half-eggs. This work demonstrated not only the feasibility of using unfertilized eggs to give rise to viable offspring, but the power of egg cytoplasm to remodel nuclei.

Similar experiments were conducted at the same time by Roger Prather at the Wisconsin Laboratory of Neal First. Unfertilized oocytes were tested as suitable recipients for blastomere nuclei. A work published in 1987 described how two calves had been born from embryos reconstructed by electrofusion of morula stage blastomere nuclei with enucleated unfertilized eggs. Step one (Figure 10.4) removed the egg's chromosomes. Although this step was termed **enucleation** by the authors, strictly speaking it was chromosome removal because metaphase II eggs do not have a nucleus. After removal of the egg chromosomes, the resulting egg cytoplast was fused with nuclei from morula stage cow embryos using

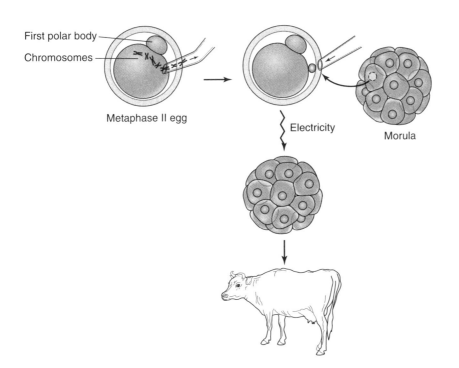

First polar body
Chromosomes

Metaphase II egg

Electricity

Morula

FIGURE 10.4
Transplantation of cow morula nuclei into unfertilized egg.

FIGURE 10.5
Endlessly cloning one cow by
recycling morula nuclei.

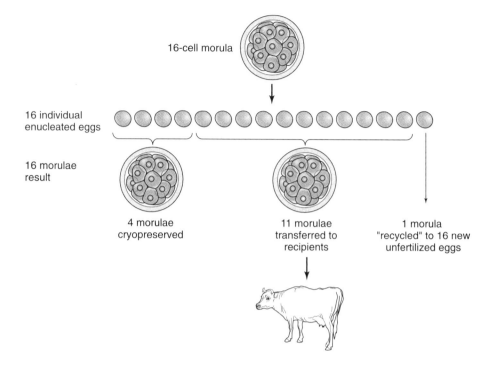

electrical pulses, as previously reported by Robl. The electrical pulses not only fused the blastomere nucleus with the egg cytoplast, they also served to activate the egg so it would initiate cleavage into blastomeres.

Although the success rates were low (less than 1% of eggs fused with morulae nuclei resulted in calves), these results raised the provocative possibility of endlessly cloning one cow (Figure 10.5). Unfertilized cow eggs are readily available from ovaries of slaughterhouse animals, thus markedly easing the logistics of needing fertilized eggs. Theoretically, each of the 16 blastomeres of a cow morula could be fused with enucleated unfertilized cow eggs, giving rise to 16 identical morulae. Some of those could be transferred to the uteri of pregnant cows for gestation, others could be preserved by freezing them in liquid nitrogen, and others could serve as the source of blastomeres for the next round of cloning. This possibility was met with enthusiasm by people wishing to produce genetically identical offspring for a variety of reasons, including livestock production, rescue of endangered species, and drug testing and other types of research. Production companies sprang up all over the world. However, the work was greeted with horror by people concerned about eugenics and the creation of an overabundance of genetically identical individuals, thus biasing the gene pool and bypassing essential processes of natural selection.

Importantly, the theoretical possibility was not realized. For reasons still not understood, second and third generations of cloned blastomeres had a markedly reduced potential to develop to offspring. The success rate was well below 1%, making the process economically unfeasible. Unraveling the biology behind this failure will provide important insights into the subtle, but highly significant, processes that underlie successful embryologic development. The work will take years and major financial resources. Research in this area is proceeding slowly.

Nonetheless, the possibility of obtaining multiple genetically identical individuals from morulae has remained attractive for research purposes, even if only one generation were possible. This work continues for research purposes in laboratories throughout the world.

Transplantation of Somatic Cell Nuclei Into Unfertilized Eggs

The ability to generate cleaving blastomeres from nuclei transplanted into unfertilized eggs of several species greatly improved the opportunity to continue to explore the developmental potential of nuclei from a variety of cell types. The goals of the continuing research included not only basic developmental biology questions related to gene expression and nuclear reprogramming, but also an interest in applying the emerging genetic manipulation techniques demonstrated in mice to larger animals. For example, as discussed in Chapter 8, mouse embryonic stem (ES) cells become incorporated into the inner cell mass (ICM) when injected into the blastocoel cavity and subsequently contribute to the tissues of the developing offspring. Because mouse ES cells can be genetically engineered in a variety of ways, including homologous recombination with altered gene sequences (see Chapter 8), it is theoretically possible to derive an endless variety of genetically altered mice. Such mice provide powerful probes into the function of specific genes and interaction with other genes. Perhaps even more provocative is the opportunity to correct mutated genes in the genetic background of the animal or to add new genes. On the one hand, this line of investigation again raises the specter of eugenics and is to be pursued with due caution; on the other hand, the amount of suffering that could be alleviated by correcting known genetic defects cannot be ignored.

In addition, there is the possibility of genetically engineering animals to express proteins of therapeutic value. One example of this is the expression of tissue plasminogen activator (TPA) in the milk of goats. TPA has been shown to improve the outcome of people suffering heart attacks if administered quickly. The gene for human TPA has been added to the germ line of some goats by injecting DNA constructs containing the gene into the pronuclei of fertilized goat eggs (as described in the sidebar of Chapter 5). Expression of TPA in goat milk apparently has no side effects to the animal. The process of deriving such an animal through transgenic technology is expensive and lengthy. Once such an animal is obtained, there would be great advantage in being able to clone her from her somatic cells. Only in this way would exact genetic replicas of her be possible.

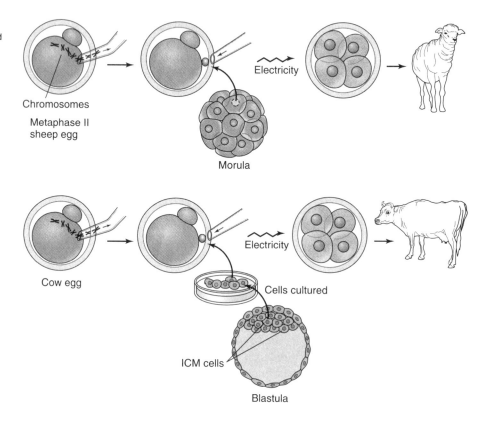

FIGURE 10.6
Sheep and cow offspring produced from morula and inner cell mass (ICM) cell nuclei.

Chromosomes

Metaphase II
sheep egg

Electricity

Morula

Cow egg

Electricity

Cells cultured

ICM cells

Blastula

By the late 1980s, groups studying nuclear transplantation began to explore the ability of ICM cells fused with unfertilized eggs to direct blastomere cleavage. In 1989, Smith and Wilmut reported the birth of lambs developed from embryos reconstituted from either 16-cell blastomeres or ICM cells fused with unfertilized sheep eggs (Figure 10.6). This was one step closer to cloning an animal from a somatic cell.

Two research teams (G. B. Anderson and Modlinski and Illmensee) attempted to clone mice from embryos reconstructed by transplantation of mouse ES cell nuclei into unfertilized mouse eggs in the early 1990s, but they failed to produce live offspring. This work reinforced the concept that only early cleaving blastomere nuclei were dedifferentiated enough to fully support embryonic development and that although cultured ES cells could contribute to most tissues when injected into blastocysts, they were not totipotent. It appeared as though there was a reduction in totipotency when ICM cells were adapted to long-term culture as ES cells.

In 1993, Sims and First reported the first calves born from embryos reconstituted from cultured ICM cells fused to unfertilized cow eggs. The ICM cells were cultured for 1 to 4 weeks at the time of nuclear transfer (Figure 10.6). An established line of ES cells had not been derived from cow blastocysts. Nonetheless, this landmark experiment proved that it may be possible to obtain offspring from ICM cells cultured long enough to derive at least a few cells with new gene inserts through homologous recombination.

This work was followed in sheep by Keith Campbell and associates, working in the laboratory of Ian Wilmut. This team derived cell lines not from blastocysts but from fetal sheep tissues. As had been demonstrated with the cow ICM cells, this team reported in 1996 that it was possible to obtain live offspring from embryos constructed by transplanting nuclei from cultured fetal cells.

Then came Dolly. In 1997, Ian Wilmut and associates reported the birth of a ewe from an embryo constructed from the nucleus of an adult sheep breast cell line that had been in culture for several generations. The prediction that "cloning by simple nuclear transfer is biologically impossible" had been resoundingly disproved.

Animal Cloning

In 1997, a series of experiments undertaken by Ian Wilmut and Keith Campbell in Edinburgh culminated in the extraordinary birth of Dolly, a white-faced Finn Dorset lamb born to a Scottish Blackface ewe (Figure 10.7). The birth of Dolly rocked science, bioethics, and venture capitalists. For the first time, and boldly in the face of prevailing scientific dogma, a mammalian egg had reprogrammed the chromosomes in an adult somatic cell that then directed embryonic and fetal development of a lamb. The adult somatic cell nucleus was that of a primary line of mammary cells recovered from the mammary gland of a 6-year-old ewe in her last trimester of pregnancy.

Ian Wilmut's quest to develop strategies for reprogramming adult nuclei began many years before Dolly, as described earlier in this chapter. His research team had focused on the concept that successful reprogramming required synchronization of the cell cycle of the transplanted nucleus with the cell cycle of the egg undergoing activation. Because the egg is arrested in mid-meiosis, it was difficult to predict whether a nucleus at early G1 or arrested in late G2 was the better cell cycle match. The team tried both and had better success with nuclei arrested in early G1, a circumstance brought about by removing serum from the cells, a well-known laboratory method for putting cells into a state of quiescence. Extended periods of culture in the absence of serum growth factors cause many cell types to exit the cell cycle, a state termed G0.

The 1997 report included lambs born from fetal and embryonic nuclei as well. The process was inefficient with all cell types, including the embryonic nuclei, thought at the time to be the most likely to succeed. The laboratory methods used had been developed during several prior years of research. They highlight the concerns about laboratory culture media and culture conditions because eggs transplanted with somatic cell nuclei were embedded in agar and surgically transferred

FIGURE 10.7
Dolly, the first cloned mammal. (*Source:* Wilmut, L., et al. Nature. 1997; 385: 810-813. Used with permission of Nature Publishing Group and Jim McWhir, Roslin Institute.)

to the oviducts of ewes for "incubation" for 5 to 7 days. The oviducts of the ewes had been surgically ligated to prevent the nuclear transplanted eggs from migrating into the uterus.

In work reported by Campbell and Wilmut in 1996, in which lambs were successfully developed from transplanted nuclei of sheep ES cells, approximately 15% of the nuclear transplanted eggs developed to morulae and blastocysts in the sheep oviduct. It is interesting to note that in the ensuing years, this percentage development has not changed significantly. Of the morulae and blastocysts that were recovered, 50 were transferred into the uteri of recipient ewes for gestation, five lambs were born, but only two survived past 10 days.

Although discouragingly inefficient, the work paved the way for the attempts to remodel nuclei from adult somatic cells that resulted in the birth of Dolly.

Although mice have been the mainstay of animal research for several decades, successful cloning of mice was not reported until 1998. A research team at the University of Hawaii in the laboratory of Yihiro Yanagamachi cloned mice from granulosa cell nuclei, and Jim Robl's team at the University of Massachusetts cloned cows from fetal fibroblasts that had been rendered transgenic by homologous recombination with a DNA construct containing a gene that encoded a protein easily distinguishable from cow proteins. These breakthroughs confirmed the feasibility of carrying out specific genetic manipulations in cattle and opened the possibility of engineering livestock to produce therapeutics in large quantities, such as in milk. In 1999, live goats were produced from embryos constructed from fetal fibroblasts, and in 2000, pigs were produced by procedures similar to the goat. Thus as of this writing, offspring have been cloned from somatic cell nuclei from cows, sheep, goats, pigs, mice, rabbits, and cats.

Importantly, however, the percentage of live offspring from reconstructed embryos remains low, on the order of 1 or 2%. In addition, some die within the first few weeks of life. The reasons for this are not clear but do not appear to be the result of chromosomal anomalies because the offspring studied, including those that died, seem to have a normal karyotype. This suggests that the deaths resulted from failures in epigenetic controls, such as those involved in imprinting.

Nonetheless, the cloning successes have demonstrated the ability of eggs to dedifferentiate somatic cells into totipotent embryonic cells. This has established a fundamentally new biologic principle. Whether or not animal cloning can be made more efficient and healthy for the offspring, the power of the discovery to derive pluripotent stem cells for therapeutic purposes has been firmly established.

In 2004, a team of South Korean researchers reported the first successful pluripotent stem cell line derived by nuclear transplant into human eggs. The following year, they reported the successful derivation of 11 more pluripotent human stem cell lines, this time from patients in need of stem cell therapy. Immediately after the appearance of the 2005 report, however, the research was challenged and quickly proven to be false. Thus, as of this writing, no human stem cells have been produced after nuclear transplantation. This is an area of active research and may require the discovery of fundamentally new information about the nature of human eggs.

Challenges in Cloning Human Pluripotent Stem Cells

As discussed in the preceding chapters, **nuclear reprogramming** and egg activation must both be accomplished to bring about the self-sustained cleavages of early blastomeres derived from transplantation of somatic cell nuclei into enucleated human eggs (see sidebar). Until methods are developed to derive pluripotent stem cell lines directly from blastomeres, stem cells need to be derived from ICM cells,

Nuclear reprogramming: the process of changing gene expression from a differentiated cell state to an undifferentiated cell state to restore pluripotency.

Somatic Cell Nuclear Transfer in Humans
Pronuclear and Early Embryonic Development

Jose Cibelli completed veterinary school in Argentina. Rather than immediately entering veterinary practice, he decided to pursue his interest in cloning cattle by enrolling in graduate school in the Animal Science program at the University of Massachusetts. He married his high school sweetheart and convinced her to travel to Massachusetts with him. It was inevitable that his interests would lead him to the laboratory of Jim Robl.

Jose honed his microscopy and micromanipulation skills by conducting experiments with both rabbit and cow eggs. In 1998, he published the studies that announced the birth of the first transgenic calves. The low efficiency of production of offspring and the relatively high frequency of perinatal death were discouraging, but Jose had already realized that the real benefit from nuclear transfer embryos was the potential to derive stem cells for therapeutic purposes.

During his graduate education, a small biotechnology company was formed. The patents that arose from the nuclear transfer experiments had attracted investors in the technology. Advanced Cell Technology was formed to facilitate continuing the work with investor financing. The goal was to develop the methods for deriving human stem cells from somatic cell nuclei (such as skin fibroblasts) transplanted into enucleated eggs. The logistical problems inherent in acquiring human eggs for research purposes were staggering, and many of the methods of enucleation, nuclear transplantation, and egg activation had not been tested in humans.

Advanced Cell Technology was acquired by Michael West, a scientist with experience in running biotechnology companies. West had previously founded Geron Corporation, a California biotechnology company focused on regenerative medicine. Dr. West had conducted seminal experiments on the enzyme telomerase, which maintains the ends of chromosomes. Shortening of the stretch of nucleic acids at the ends of chromosomes is thought to play a role in cellular aging. Dr. West's passion to develop human therapeutic stem cells matched that of Jose Cibelli.

The for-profit nature of the company brought sharp criticism from critics of the proposed work, to conduct nuclear transplantation into human eggs, which would then be activated to initiate self-sustained cleavages to the blastocyst stage. The initial steps were akin to cloning a human. The company and its scientists buffeted the public outcry and compiled an ethics advisory board to provide outside guidance for recruiting women to donate eggs for research. Sharp criticism was then leveled at the ethics board as well. Congress threatened to outlaw all forms of human nuclear transplantation into eggs for the purposes of deriving blastocysts for research.

Nonetheless, carefully following the lengthy guidelines of the ethics board, Jose Cibelli and his colleagues were able to stimulate a few human eggs into initiating cleavage divisions, both without and with nuclear transplantation. The work was published in 2001. A total of 71 eggs was recovered from the ovaries of seven women research volunteers. Twenty-two were activated parthenogenically to test the protocols for artificial egg activation. Ninety percent cleaved, 30% of which developed to at least the early blastocyst stage. This result predicted that it would be possible to derive human ES cells from parthenogenically activated blastocysts. If such ES cells were proven to be histocompatible with the woman from whom they were derived, this in itself offered an opportunity to develop stem cell therapy for ovulating women.

In addition, 11 eggs underwent nuclear transplantation with cultured fibroblast nuclei before they were activated. Although some nuclear remodeling was evident, none of the eggs cleaved. Eight eggs underwent nuclear transplantation with granulosa cell nuclei from the egg donors. Three of those eight cleaved beyond two cells. Although the numbers were small, the road to human nuclear transplant stem cells had begun.

in a manner similar to that of the ES cells of the mouse. Thus for the foreseeable future, human stem cells derived from nuclear transplantation depend on the ability to activate enucleated eggs with transplanted nuclei in a manner analogous to animal cloning.

The cell cycle stage of the somatic cell nucleus being transferred has been found to play a role in successful subsequent cleavages. Because the activated egg is completing M phase and immediately entering G1, the somatic cell nuclei are most synchronous if they are arrested in G1 of their cell cycle. Several methods are available to bring about a G1 arrest of tissue culture cells, including withdrawing the growth factors present in the serum that drives cell division. If this is accomplished, the G1 phase of the activated egg may be too short to allow reprogramming of the somatic cell nucleus, as well as recruitment into S phase, as directed by the cell cycle proteins. This is an area of active investigation.

Taken together, work to date has revealed that nuclear reprogramming involves decreased expression of cell-type specific genes, for example, hemoglobin, and increased ease of expression of a broader range of genes, especially those known to be expressed in the ICM, such as *OCT4*. Importantly, some gene products function to inhibit the expression of other genes. It is possible that a fundamental feature of reprogramming a somatic cell nucleus to support embryonic development is broad suppression of all genes leading to cell differentiation coupled with broad expression of all genes involved in DNA replication and cell division. Given that this can be accomplished, the next requirement would be to invoke the systematic expression of groups of genes in the correct order to bring about cell differentiation.

Summary

Early studies with zygotes and early cleaving blastomeres suggested that the potential for mammalian cells to direct embryonic development was restricted within the first few cleavage divisions, in contrast to amphibian cells. By the late 1980s, scientists had begun to explore the feasibility of constructing embryos from unfertilized eggs rather than from zygotes. Successes with this approach accelerated research, particularly with large animals, because of the ready availability of cow and sheep eggs from slaughterhouses. The theoretical possibility of endlessly cloning one cow embryo met with such reduced success at each round of cloning, for reasons still not explained, that the approach was abandoned. Each year throughout the early 1990s brought improved success with reconstructing embryos from unfertilized eggs and cells at increasingly advanced stages of differentiation, until finally Dolly the sheep was born, reconstructed with the nucleus of a long-term tissue culture cell line. This opened the door to cloning all species from adult somatic cells and launched the ethical debates about eugenics that rage today.

Despite aggressive studies in laboratories all over the world, the success of animal cloning has not substantially improved. A central problem is the efficiency of reprogramming the transplanted nucleus. It swells enormously upon being transplanted into the egg, and presumably the dramatic increase in open scaffolding has an effect on ease of gene expression. But size alone is not sufficient. Specific modifications to the chromatin are clearly needed to bring about a fully dedifferentiated state. A recently reported candidate for chromatin remodeling is nucleoplasmin, the protein helper (chaperone) that guides proteins such as histones to their positions. Improvements in nuclear remodeling may require pretreatment of nuclei to initiate nuclear remodeling prior to transplantation. Much research is needed.

Review Questions (True or False)

1. In contrast to amphibian chromosomes, mammalian chromosomes cannot be remodeled to support embryonic development.
2. Animal cloning involves transplantation of the entire set of chromosomes from a somatic cell into an egg whose chromosomes have been removed.
3. An egg with a transplanted nucleus that contains a full set of chromosomes could lead to an offspring, but only if transplanted into a uterus.
4. The first cloned mammal was Dolly the sheep, born in 1997.
5. The common laboratory method for cloning mammals that employs cell fusion to transplant somatic cell chromosomes into a chromosome-free egg also transplants the somatic cell mitochondria.
6. Animal cloning provides a new opportunity for increasing the number of animals with desirable genetic traits.
7. The efficiency of cloning animals remains low despite global research efforts to improve it.
8. Transplanting human somatic cell nuclei into chromosome-free human eggs holds the promise of deriving pluripotent stem cells genetically identical to the patient in need.

ADDITIONAL READINGS

Baquisi, A., Behboodi, E., Melican, D. T., Pollock, J. S., Destrempes, M. M., Cammuso, C., Williams, J. L., Nims, S. D., Porter, C. A., Midura, P., Palacios, M. J., Ayres, S. L., Denniston, R. S., Hayes, M. L., Ziomek,

C. A., Meade, H. M., Godke, R. A., Gavin, W. G., Overstrom, E. W., and Echelard, Y. (1999). Production of goats by somatic cell nuclear transfer. *Nat Biotechnol* 17: 456–461.

Campbell, K. H. S., McWhir, J., Ritchie, W. A., and Wilmut, I. (1996). Sheep cloned by nuclear transfer from a cultured cell line. *Nature* 380: 64–66.

Illmensee, K., and Hoppe, P. C. (1981). Nuclear transplantation in *Mus musculus*: developmental potential of nuclei from preimplantation embryos. *Cell* 23: 9–18.

Kuroiwa, Y., Kasinathan, P., Choi, Y. J., Naeem, R., Tomizuka, K., Sullivan, E. J., Knott, J. G., Duteau, A., Goldsby, R. A., Osborne, B. A., Ishida, I., and Robl, J. M. (2002). Cloned transchromosomic calves producing human immunoglobulin. *Nat Biotechnol* 20: 889–894.

McGrath, J., and Solter, D. (1984). Inability of mouse blastomere nuclei transferred to enucleated zygotes to support development in vitro. *Science* 226: 1317–1319.

Modlinski, J. A. (1981). The fate of ICM and trophectoderm nuclei transplanted to fertilized mouse eggs. *Nature* 292: 342–343.

Sims, M., and First, N. L. (1993). Production of calves by transfer of nuclei from cultured inner cell mass cells. *Proc Natl Acad Sci USA* 90: 6143–6147.

Tamada, H., Thuan, N. V., Reed, P., Nelson, D., Katoku-Kikyo, N., Wudel, J., Wakayama, T., and Kikyo, N. (2006). Chromatin deconsation and nuclear reprogramming by nucleoplasmin. *Mol Cell Biol* 26: 1259-1271.

Willadsen, S. M. (1986). Nuclear transplantation in sheep embryos. *Nature* 320: 63–65.

Willmut, I., Schnieke, A. E., McShir, J., Kind, A. J., and Campbell, K. H. S. (1997). Viable offspring derived from fetal and adult mammalian cells. *Nature* 385: 810–814.

Yanagimachi, R. (2002). Cloning: experience from the mouse and other animals. *Mol Cell Endocrinol* 187: 241–248.

PART

IV

Human Stem Cells

CHAPTER 11

Adult and Fetal Stem Cells

I have not failed. I've just found 10,000 ways that won't work.

Thomas Alva Edison

OVERVIEW

As discussed in previous chapters, stem cells are capable of unlimited cycles of growth and division into genetically faithful daughter cells and can be induced to differentiate into a wide variety of cell types. It has been known for some time that there are multiple classes of stem cells, some of which are already committed to a tissue type, e.g., skin stem cells. The nomenclature is imprecise, and stem cells are generally classified as either "adult," "fetal," or "embryonic," as described in Chapter 1. This classification refers simply to their source and does not accurately reflect their level of commitment or their differentiation potential. In many instances, commitment and differentiation potential are not known with certainty. As a general rule, adult stem cells are thought to be more committed than fetal stem cells, which are in turn more committed than embryonic stem (ES) cells. As the stem cell field matures, more accurate nomenclature will undoubtedly develop.

The overarching consideration for stem cells is how to get them from the laboratory to clinical treatments. Much needs to be done. This chapter and Chapter 12 outline what is known and what is not known to accomplish this.

Adult Stem Cells

The term **adult stem cells** refers to the self-renewing population of cells in those adult tissues capable of routine self-renewal. Hair, skin, blood, lining of the intestine, liver, and male germ cells are examples of tissues that are capable of self-renewal. It is important to point out that not only are such tissues capable of replacing cells lost due to insults to the tissues, such as trauma, but self-renewal is an innate feature of their role in the body. Hence, part of their tissue organization is the sequestering of a population, albeit very small, of cells with the ability to divide at whatever rate is required to generate sufficient numbers of cells to replace losses. The process is complicated and has multiple stages.

Two terms are commonly used to describe the stages from stem cell to differentiated adult cell. One is **progenitor,** used to designate a stem cell fully committed to a general type of adult cell. The other is **precursor**, used to designate a stage of differentiation close to the final adult cell. The generation of sperm and blood cells are well-characterized examples of the stages required to differentiate fully functioning adult cells from stem cells.

Sperm Production

Baby boys are born with a population of sperm stem cells, **spermatogonia,** that are sequestered within tubules inside the testes. They remain sequestered and quiescent until puberty heralds the hormonal stimulation from the pituitary gland that awakens the testosterone-producing cells in the testis. A cascade of cellular signals, not fully understood, stimulates the spermatogonia to initiate the cycles of cell division that persist in the adult male for his lifetime. In a beautifully orchestrated sequence (Figure 11.1), one of the daughter spermatogonium enters the multistep pathway to mature sperm. It goes through several stages of chromatin reorganization and enters meiosis, ultimately giving rise to four round immature spermatocytes, termed *spermatids*. They undergo several more stages of chromatin reorganization, eventually growing a long tail and condensing their DNA into a tightly packed silent coil within the sperm head. The cells remain associated with each other as they differentiate, expanding toward the center of the tubule until the sperm are swept from the tip of the rising pyramid to exit the testis.

In this example, the spermatogonia are the progenitor cells and the spermatids are precursor cells. The stem cells giving rise to the spermatogonia were the primordial germ cells that migrated to the gonadal ridge to form the testis, described in Chapter 9. Although the primordial germ cells are committed to becoming sperm or eggs, the commitment can be altered because cells from germ cell tumors, such as embryonal carcinomas, can be induced to form several types of cells in laboratory culture, as described in Chapter 9. In contrast, spermatogonia are known to give rise only to sperm. Laboratory systems for culturing spermatogonia have only recently been developed, however, by Australian scientist Suzanne Hasthorpe, so perhaps their full developmental potential remains to be discovered.

Bone Marrow Stem Cells

Diane Kraus and Neil Theise and their colleagues provided evidence in 2001 that a single bone marrow stem cell transplanted into a lethally irradiated

169

FIGURE 11.1
Spermatogenesis. Spermatogonia are the stem cell progenitors to sperm, and spermatocytes are the haploid precursor cells.

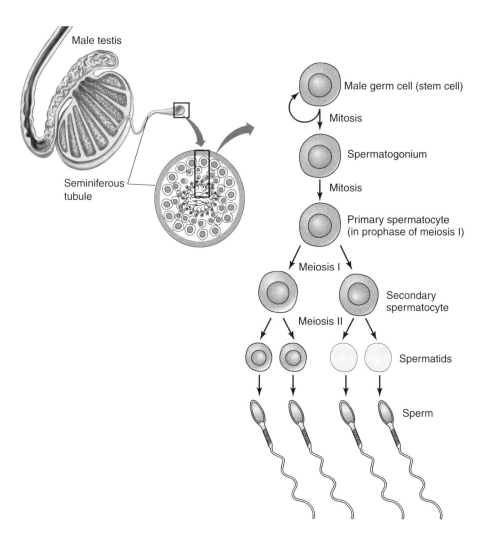

mouse not only gave rise to sufficient bone marrow cells to repopulate the animal's entire blood cell repertoire, but also contributed to a variety of other tissues. The technical challenge to such experiments is identifying the daughter cells from the original bone marrow stem cell. These investigators transferred one male bone marrow stem cell into each of 20 female mice whose bone marrow cells had been killed by irradiation. At 12 weeks, five females had evidence of normal bone marrow production and their blood cells had a Y chromosome. At 11 months, their bone marrow was harvested and transferred into a second set of five female recipients to test the pluripotency of the daughter cells of the original single bone marrow stem cell. Four months later, three of the five exhibited evidence of normal bone marrow function, producing cells with a

Y chromosome. These extraordinary results confirmed the pluripotency of the original male bone marrow stem cell.

In addition to blood cells, cells bearing a Y chromosome were detected among the epithelial cells of lung, esophagus, stomach, intestine, bile duct of the liver, and skin. The evidence is convincing that progeny of the original single bone marrow cell were found in all such tissues, but whether or not they had actually differentiated into each cell type is more difficult to judge, especially in light of a growing appreciation of the important role that tissue-specific macrophages play in normal development and function of most tissues.

These results extended prior reports of bone marrow stem cells giving rise to cells in the liver. The ability of stem cells from adult tissues to develop not only into cells of their tissue of origin but also into cells of other tissues is an active area of investigation in many laboratories throughout the world. The term **transdifferentiation** is applied to the property of stem cells to change their fate. Given the inherent plasticity and pluripotency of bone marrow stem cells, they are likely candidates for transdifferentiation if it occurs.

One suggestion for a more accurate term for adult stem cells has been **multipotent adult progenitor cells** (MAPCs). This term includes the emerging notion of transdifferentiation. Based on embryology (see Chapter 9), one line of thinking about adult stem cells has been that they would stay faithful to their tissue of origin, or at the least to their tissue of embryonic origin. For example, stem cells from adult bone marrow obviously give rise to all the blood cell lineages (see sidebar). In addition, they could perhaps give rise to other cell types from tissues of mesodermal origin but not from ectodermal or endodermal origin. At least for mice and rats, this long-held view has now been challenged by the reports of bone marrow cells giving rise to liver cells. Additional challenges have been reported by Catherine Verfaillie at the University of Minnesota. In a series of elegant experiments, her research team isolated mouse MAPCs (mMAPCs) from mouse bone marrow and first demonstrated that a single cell was capable of continually dividing into a large number of identical daughter cells (termed **clonal expansion**). In addition, when those cells were exposed to appropriate growth factors in laboratory culture systems, they developed the characteristics of both neuronal cells (ectodermal) and liver cells (endodermal).

Clonal expansion: one cell giving rise to millions of identical cells.

Moreover, single mMAPCs were injected into mouse blastocysts (see sidebar, Chapter 8) to generate chimeric offspring. If the single mMAPC could divide in the inner cell mass along with the blastocyst inner cell mass cells, then in theory it could contribute to a variety of tissues in offspring along with the inner cell mass cells themselves. The single mMAPC injected into the blastocysts contained a reporter gene to allow it to be identified. The injected blastocysts were returned to the uteri of recipients for gestation and birth. The tissues of such offspring were examined for evidence of the single injected mMAPC. Although only a small percentage of offspring were successful chimeras, those that were successful contained mMAPCs in brain, retina, liver, intestine, kidney, spleen, bone marrow, and skin (Figure 11.2). These findings suggest either that the bone marrow mMAPCs were

Hematopoietic and Cord Blood Stem Cells

John Freeman Loutit, a British hematologist, was born in 1910. At age 33 he developed the citric acid-glucose solution still used today to bank blood for transfusions. In 1956, with his colleagues D. W. H. Barnes, J. L. Hamerton, and C. E. Ford, he reported that lethally irradiated mice died from bone marrow failure and that they could be rescued by infusion with nonirradiated cells from bone marrow. Five years later, Canadian researchers James Till and Ernest McCulloch reported that transplanted bone marrow cells established obvious easy-to-count colonies in the spleens of the rescued mice. In 1963, using newly developed techniques to spread and characterize chromosomes, they determined that each of the new colonies in the spleen were clonal because they had arisen from a single transplanted cell.

The spleen colony assay became a widely used tool to investigate the nature of individual transplanted bone marrow cells. It was known that bone marrow cells were precursors to cells in circulation in peripheral blood and that there were many steps involved in cellular differentiation that spanned days to weeks to accomplish. Spleen colonies that appeared, for example, 6 days after bone marrow transplantation were assumed to arise from cells more fully differentiated than those giving rise to spleen colonies 12 days after the transplant.

By the mid-1970s, the assay had been used to determine that bone marrow contained stem cells that gave rise to the entire red blood cell lineage and other stem cells that gave rise to the myeloid branch of white blood cells. Awareness that specific proteins appeared on the surface of the cells at each step of differentiation led to the development of methods to separate out and characterize specific cell populations. In 1982, American scientists Cheryl Whitlock and Owen Witte were able to culture antibody-producing cells (B lymphocytes) and their precursors for up to a year in the laboratory, thus allowing more detailed studies of stages of differentiation. A few years later, American scientist Irving Weissman reported studies of immune-responsive lymphocytes (T lymphocytes), thus paving the way for characterization of the steps of cell differentiation of all the major branches of blood cells.

Was there a single pluripotent bone marrow stem cell that gave rise to all red and white blood cell types? Or was there a pluripotent stem cell for each of the major lineages? Irv Weissman reasoned that diluting down the numbers of bone marrow stem cells transplanted, theoretically to one, would answer these questions. Cell death associated with the technicalities of the experiment eliminated the possibility of such a direct approach, but progress in enriching for pluripotent stem cells with specific traits revealed, in 1990, that, indeed, a single stem cell gives rise to all blood cell types for life. It constitutes approximately 0.04% of bone marrow cells.

Edward A. (Ted) Boyse was an immunologist at Memorial-Sloan Kettering Institute in New York City in 1982 when he proposed the idea that cord blood stem cells might be a suitable replacement for bone marrow stem cells if a bone marrow tissue match were not obtainable. The idea caught the attention of a young clinician-scientist, Harold Broxmeyer, who conducted some analyses of cord blood stem cells to determine their responses to laboratory culture conditions.

In 1988, the plight of a 5-year-old American boy living in Paris and suffering from Fanconi anemia, a fatal disease, came to the attention of Hal Broxmeyer, who had taken a faculty position at the University of Indiana. The patient's mother was pregnant, expecting to deliver in Durham, North Carolina. Gordon Douglas, a member of the cord blood study team, flew to Durham to collect the cord blood from the baby girl who was a tissue match for her older brother. It was sent to Broxmeyer's laboratory that isolated the cord blood stem cells and shipped them to Paris. The child underwent a successful cell infusion and is alive and well today. This heralded a new era of stem cell therapy.

An estimated 5,000 umbilical cord blood transplants have been performed in the ensuing 17 years, many at the University of Indiana clinic started by Hal Broxmeyer. The work has not only made life-saving transplants available to many who were not candidates for bone marrow transplants, but comparisons between the dynamics of blood cell replacement by cord blood stem cells with bone marrow stem cells has provided valuable insights into the many steps in blood cell differentiation.

Banking the cord blood cells of a newborn for future personal use is an obvious possibility but has been slow to become common practice because of logistical difficulties in acquisition and long-term banking. Perhaps with renewed interest in the pluripotent nature of cord blood cells, the value of personal banking may become more compelling.

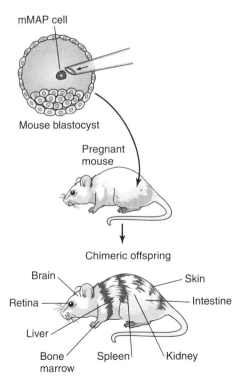

FIGURE 11.2
Tissue distribution of single mouse MAPC (mMAP cell) injected into a mouse blastocyst.

FIGURE 11.3

Tissue distribution of mouse
MAPCs (mMAP cells) injected into
the tail vein of an adult mouse.

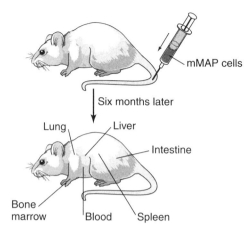

intrinsically capable of contributing to tissues of mesodermal, endodermal, and ectodermal origin or that they acquired that ability during their time in laboratory culture.

As an additional test of the potential for the mMAPCs to contribute to adult tissues, they were injected into blood veins of adult mice to determine whether they could establish themselves in any tissues. After a few months, tissues of the injected mice were examined for evidence of the mMAPCs; they were found in blood, bone marrow, and spleen as well as lung, liver, and intestine (Figure 11.3). These tissues are known to have their own population of stem cells. Importantly, the mMAPCs were not found in skeletal muscle, heart muscle, brain, or kidney, all tissues known to be deficient in their own populations of stem cells. These findings suggest that there are rigorous mechanisms for controlling stem cell populations in organs that may need to be overcome for stem cell therapy to be clinically effective. It is also important to point out that lung, liver, and intestine have tissue-specific macrophages that may partially explain the distribution of the injected mMAPCs.

These findings about adult bone marrow stem cells are being tested in other laboratories. Alternative explanations, such as fusion with embryonic cells or differentiation into tissue specific macrophages, have already been put forth by other experts in the field. This is an active area of investigation, with new findings reported weekly.

The possibility that bone marrow stem cells can differentiate into heart muscle cells has received considerable attention since it was first reported in a mouse model in 2001 by the laboratories of Piero Anversa, New York Medical College, and Donald Orlic, National Institutes of Health. This transdifferentiation was refuted in 2004 by two research teams, Balsam and colleagues at Stanford University and Murry and colleagues at the University of Washington. There is no debate about the innate ability of bone marrow stem cells to home to the site of damaged heart cells. What is debated is whether or not the bone marrow stem cells differentiate into mature blood cells or transdifferentiate into heart muscle cells, termed **cardiomyocytes.** There is some evidence that the bone marrow stem cells fused with the cardiomyocytes. It is important to note that all research teams report the bone marrow stem cells improve the function of damaged hearts, independent of their ascribed cell type.

More recent studies reported in 2005 by a Boston team of scientists lead by Doug Losordo have confirmed that a pluripotent human bone marrow stem cell is also capable of clonal expansion from a single cell. Ten clones with robust tissue culture characteristics were obtained from each of three donated samples of bone marrow. Once established in culture, the cells were exposed to culture conditions used to induce differentiation of human embryonic stem (ES) cells. The bone marrow stem cells responded in a manner similar to the ES cells, giving rise to cells with characteristics of blood vessels, smooth muscle cells, neurons, liver, and heart muscle. To test the response of the cloned bone marrow stem cells to cardiac insult, they were delivered to damaged hearts in a rat model. Examination of the hearts revealed evidence of fusion between the human cells and the rat heart cells as well as expression of some heart muscle cell genes by the human cells.

Whether human bone marrow stem cells will ultimately be proven to be capable of differentiation into functioning heart muscle cells, or neurons, or liver cells, or other cell types requires more research. The laboratory approaches and techniques developed to address these issues will also have broad applicability to ES cell studies, as described in Chapter 12. Research to date, however, demonstrates a heretofore unappreciated plasticity of human bone marrow stem cells with respect to the potential to express genes known to be active in a wide variety of fully differentiated adult tissues. This information alone bodes well for a stem cell approach to repair of tissues deficient in stem cells of their own.

Cord Blood Stem Cells

The first therapeutic use of cord blood stem cells was reported in 1988 (see sidebar). In 1992, the New York Blood Center's Placental Blood Program was established. It now has 13,000 donations available for transplantation into children who need hematopoietic stem cells. It is rarely used for adults because one cord blood collection is estimated to yield only a few million stem cells, sufficient for a small child but too few for an adult. A research goal is, therefore, to discover ways to culture bone marrow stem cells to increase their numbers.

Another area of active research is understanding the nature of cord blood stem cells, particularly with respect to pluripotency and to characteristics in common with bone marrow stem cells. The work is in early stages because of the lack of routine laboratory methods for expansion of cord blood cells and fractionation into specific cell types. This problem touches on a fundamental problem with stem cell biology: whether or not they should be dividing. Pluripotent stem cells divide relentlessly only during embryonic and fetal development. Stem cells in adult tissues, by definition, divide only when called upon to repair damage. Loss of control over cell division, especially in bone marrow, leads to cancers, such as leukemias and lymphomas. Hence, expansion of any adult stem cell requires first overcoming the innate block to cell division in the absence of specific signals needed to recruit them.

It is possible that cord blood cells contain stem cells with less stringent controls on cell division and that it may become possible to obtain greater numbers of pluripotent stem cells from the umbilical cords of newborns than from the bone marrow of adults. The work is ongoing.

Neural Stem Cells

Studies in rodents and nonhuman primates have revealed that the adult brain has at least two repositories for stem cells: the **subventricular zone** and the **subgranular zone** of the **hippocampus** (see Figures 13.1 and 13.2). The subventricular zone supplies new neurons for the olfactory bulb, which has a higher turnover rate for neurons than any other area of the brain. This relates to the continuing need to provide nerve connections to the thousands of receptors that constitute our sense of smell.

Despite the existence of brain stem cells, repair of damaged brain tissue is largely nonexistent, for reasons unknown. Perhaps it relates to the greater importance of mechanisms to prevent cell division within the brain whose size is tightly constrained by the inflexible volume of the skull. This need may have resulted in more stringent controls on brain stem cells than on stem cells in other adult tissues.

During development of the brain (see Chapter 9), two internal fluid-filled areas are formed, the ventricles. The subventricular zone is the area directly underneath the ventricles. Many years of rodent studies have revealed new neurons arise within the subventricular zone and migrate to the olfactory bulb. These findings were confirmed in human brains in 2004 by Canadian scientists Andreanne Bedard and Andre Parent using a variety of techniques to identify immature and mature neurons. In this regard, it is important to note that loss or dysfunction of sense of smell has for many years been associated with several neurologic disorders, such as schizophrenia and Alzheimer, Huntington, and Parkinson diseases. This suggests that dysfunction in the generation of new neurons may play a role in these diseases, providing further support that the introduction of new neurons via stem cell therapy may alleviate these devastating neurologic disorders. Research to learn to recover and expand neuronal stem cells from the subventricular zone and the subgranular zones of the hippocampus is in early stages and will undoubtedly provide fundamentally important information about controls on the development of a variety of the many types of specialized neurons necessary to carry out brain functions.

Fetal Stem Cells

Three sources of fetal stem cells have been investigated: trophoblast stem (TS) cells, primordial germ cells, and fetal tissue stem cells. As with some other types of stem cells, the principal difficulty in obtaining long-term self-renewing populations of TS cells was the inability to prevent them from differentiating. This is an important observation and seems to also apply to many, if not all, stem cells. Under many laboratory culture conditions, stem cells spontaneously differentiate and stop dividing. This suggests that differentiation is the default mode for cells, which, in many respects, provides a natural limitation to the size of organs and a barrier to tumor formation. Therefore, the environment that supports an uncommitted state must promote expression of those genes whose products inhibit differentiation. In tissues, the environment must also inhibit cell division unless new cells are needed, such as described for the skin injury in Chapter 2.

Therefore a fundamentally important development that led to successful laboratory culture of TS cells was the discovery, in the laboratory of Janet Rossant in Toronto, that the growth factor fibroblast growth factor 4 (which uses heparin as a cofactor) was as necessary as the fibroblast feeder layer to maintain trophoblast cells in an undifferentiated state. This has allowed the development of continually renewing cell lines of TS cells for study. Injection of TS cells into blastocysts has revealed that they contribute exclusively to trophoblastic tissues in developing mouse fetuses, thus limiting their usefulness as therapeutic cells for other tissues. Additional studies of TS cells have revealed, however, that conditioned medium from the fibroblast feeder layers substitutes for the fibroblast cells themselves. This observation opens the way to identify those factors contributed by the feeder layer that are necessary to maintain the undifferentiated state of the TS cells. Such factors may also play an important role in maintaining the undifferentiated state of other stem cells.

Circumventing the need for a feeder layer to promote division and inhibit differentiation of stem cells is important to stem cell therapy. All lines of human ES cells currently available were derived atop a feeder layer of mouse fibroblasts. Thus, the ES cells were continually exposed to mouse cells, and there is concern that those mouse cells may spontaneously express unusual pathogens, such as endogenous retroviruses. Although this concern is theoretical at this time, there is a need to prove the safety of mouse fibroblast feeder layers for stem cells being cultured for therapeutic purposes.

Stem cells from human primordial germ cells, termed *human embryonic germ* (EG) cells, were first isolated in the laboratory of John Gearhart in 1999. Their origin was the gonadal ridge (see Chapter 9) of 5- to 9-week fetuses from elective abortions. Although highly controversial, the characteristics of EG cells may provide valuable clues about maintaining the fully undifferentiated state of human ES cells from less controversial sources than leftover human embryos in fertility clinics. A comparison of their gene expression and culture characteristics with other sources of human ES cells is listed in Table 11.1. It may be an important feature of their therapeutic developmental potential that they do not grow into teratomas in immune compromised mice.

Fetal Tissue Stem Cells

As described in Chapter 9, the mesencephalic region of the developing brain is rich in neurons that express dopamine by the 10th week of development. Some

	TABLE			
11.1	**Comparison of Human Pluripotent Stem Cells**			

Characteristic	Embryonic Stem Cells from Eggs Fertilized by Sperm	Embryonic Germ Cells from Human Fetal Gonadal Ridge	Embryonic Carcinoma Cells from Human Tumors
Express stage-specific embryonic antigens (SSEA)-1, -3, -4	SSEA-3, -4	SSEA-1, -3, -4	SSEA-3, -4
Express tumor rejection antigen (TRA)-1	Yes	Yes	Yes
Express Oct-4	Yes	Yes	Yes
Express alkaline phosphatase	Yes	Yes	Yes
Express telomerase	Yes	Unknown	Yes
Factors that aid in self-renewal	Feeder cell layer plus either serum or bFGF	LIF, bFGF, and forskolin	Not well characterized
Growth characteristics in culture	Form flat loose aggregates	Form rounded multilayer clumps	Form flat loose aggregates
Can form embryoid bodies	Yes	Yes	Yes
Grow into teratomas in nude mice	Yes	No	Yes

bFGF, basic fibroblast growth factor; LIF, leukemia inhibitory factor.

success in laboratory methods to culture these cells and expand their numbers has been reported. Both in rodent models for Parkinson disease and in a few human clinical trials with parkinsonian patients, fetal neurons introduced into the brains showed alleviation of symptoms for extended periods of time. Results were not entirely consistent from trial to trial, and some uncomfortable side effects, such as increased uncontrolled movement of the extremities, appeared to be worse in some human recipients of fetal cells. These side effects coupled with the difficulty in obtaining sufficient fetal cells for routine therapies have all but eliminated human trials with fetal tissues. The positive results, however, provide support for continued efforts in cell replacement with pluripotent stem cells.

As discussed in Chapter 14, fetal pancreatic tissues have also been used to treat diabetes with promising results. As with the Parkinson's trials, lack of efficacy in all recipients, coupled with the difficulty in obtaining sufficient pancreatic tissues, has also nearly eliminated human trials, awaiting the development of more consistent insulin-producing cells from pluripotent stem cells.

Summary

The term *adult stem cell* refers to the stem cell population in normal adult tissues. Fetal stem cells are those present in the tissues of developing fetuses. It is possible that fetal stem cells are more multipotent than adult stem cells, but information is incomplete. It is also possible that fetal and/or adult stem cells can be coaxed to transdifferentiate into cells of other types of tissues than their origin, but more research is needed. An example is the refuted reports that bone marrow stem cells have the capacity to differentiate into cardiac muscle and aid in the repair of damaged hearts. The evidence that bone marrow stem cells home to damaged heart muscle and promote heart muscle strength was not refuted, but transdifferentiation into heart muscle cells was refuted.

Differentiation of adult tissue stem cells requires several steps to full maturity, as exemplified by sperm and blood cell production. Some nerves, such as olfactory neurons, also continually self-renew, providing a possible source of neuronal stem cells. Evidence that the stem cells that give rise to sperm and olfactory neuron can transdifferentiate into other tissue types is lacking. But the remarkable ability of a single bone marrow stem cell to give rise to all the types of blood cells, both red and white, indicates bone marrow stem cells may have greater transdifferentiation potential. And the embryonic precursor to the sperm stem cell, the embryonic germ cell, also exhibits multipotential characteristics in laboratory culture.

Similarly, placenta TS cells can be expanded in laboratory culture, but their ability to differentiate into all body tissues has not been demonstrated. Clinical trials with fetal insulin-producing cells and fetal brain cells have demonstrated clinical improvement in patients with diabetes and Parkinson disease, however, demonstrating the potential for stem cell therapies. Umbilical cord blood stem cells may also have widely applicable therapeutic potential if laboratory methods for expanding their numbers and differentiating them into desired cell types can be discovered.

Review Questions (True or False)
1. Bone marrow stem cells home to damaged heart muscle and improve heart function.
2. Cord blood stem cells can be used to repopulate the bone marrow, giving rise to all blood cell types.

3. TS cells and embryonic germ cells give rise to teratomas in a mouse model.
4. *Transdifferentiation* is the term used to describe the capacity for adult stem cells to differentiate into other types of tissue than their tissue of origin.
5. Scientists may discover that some diseases can be treated by adult or cord blood stem cells more effectively than by ES cells.
6. Cord blood stem cells have been shown to cure type 1 diabetes.
7. Progenitor cells are the precursors to ES cells.

ADDITIONAL READINGS

Balsam, L. B., Wagers, A. J., Christensen, J. L., Kofidis, T., Weissman, I. L., and Robbins, R. C. (2004). Haematopoietic stem cells adopt mature haematopoietic fates in ischaemic myocardium. *Nature* 428: 668–673.

Becker, A. J., McCulloch, E. A., and Till, J. E. (1963). Cytological demonstration of the clonal nature of spleen colonies derived from transplanted mouse marrow cells. *Nature* 197: 452–454.

Bedard, A., and Parent, A. (2004). Evidence of newly generated neurons in the human olfactory bulb. *Dev Brain Res* 151: 159–168.

Hasthorpe, S. (2003). Clonogenic culture of normal spermatogonia: in vitro regulation of postnatal germ cell proliferation. *Biol Reprod* 68: 1354–1360.

Jackson, K. A., et al. (2001) Regeneration of ischemic cardiac muscle and vascular endothelium by adult stem cells. *J Clin Invest* 107: 1395–1402.

Jiang, Y., Jahagirdar, B., Reinhardt, R., Schwartz, R., Keene, C., Ortiz-Gonzales, R., et al. (2002). Pluripotency of mesenchymal stem cells derived from adult marrow. *Nature* 418: 41–49.

Kirschstein, R., and Skirboll, L. R., eds. (2000). *Stem Cells: Scientific Progress and Future Research Directions*. www.nih.gov

Krause, D. S., Theise, N. D., Collector, M. I., Henegariu, O., Hwang, S., Gardner, R., Neutzel, S., and Sharkis, S. J. (2001). Multi-organ, multi-lineage engraftment by a single bone marrow-derived stem cell. *Cell* 105: 369–377.

Lundvall, O., Kokaia, Z., and Martinez-Serrano, A. (2005). Stem cell therapy for human neurodegenerative disorders—how to make it work. *Nat Med* 10(Suppl): S42–S50.

Marshak, D. R., Gardner, R. L., and Gottlieb, D. eds. (2001). *Stem Cell Biology*. New York: Cold Spring Harbor Laboratory Press.

Murry, C. E., Soonpaa, M. H., Reinecke, H., Nakajima, H., Nakajima, H. O., Rubart, M., Pasumarthi, K. B., Virag, J. I., Bartelmez, S. H., Poppa, V., Bradford, G., Dowell, J. D., Williams, D. A., and Field, L. J. (2004). Haematopoietic stem cells do not transdifferentiate into cardiac myocytes in myocardial infarcts. *Nature* 428: 664–668.

Niwa, H., Miyazaki, J., and Smith, A. G. (2002). Quantitative expression of Oct-3/4 defines differentiation, dedifferentiation or self-renewal of ES cells. *Nat Genet* 24: 328–330.

Orlic, D., Kajstura, J., Chimenti, S., Jakoniuk, I., Anderson, S. M., Li, B., Pickel, J., Nadal-Ginard, B., Bodine, D. M., Leri, A., and Anversa, P. (2001). Bone marrow cells regenerate infracted myocardium. *Nature* 401: 701–705.

Petersen, B. E., Bowen, W. C., Patrene, K. D., Mars, W. M., Sullivan, A. K., Murase, N., Boggs, S. S., Greenberger, J. S., and Goff, J. P. (1999). Bone marrow as a potential source of hepatic oval cells. *Science* 284: 1168–1170.

Shizuru, J. A., Negrin, R. S., and Weissman, I. L. (2005). Hematopoietic stem and progenitor cells: clinical and preclinical regeneration of the hematolymphoid system. *Annu Rev Med* 56: 509–538.

Spangrude, G. J., Heimfeld, S., and Weissman, I. L. (1988). Purification and characterization of mouse hematopoietic stem cells. *Science* 241: 58–62.

Theunissen, K., and Verfaillie, C. M. (2005). A multifactorial analysis of umbilical cord blood, adult bone marrow and mobilized peripheral blood progenitors using the improved ML-IC assay. *Exp Hematol* 33: 165–172.

Yoon, Y.-S., Wecker, A., Heyd, L., Park, J.-S., Tkebuchava, T., Kusano, K., Hanley, A., Scadova, H., Qin, G., Cha, D.-H., Johnson, K. L., Aikawa, R., Asahara, T., and Losordo, D. W. (2005). Clonally expanded novel multipotent stem cells from human bone marrow regenerate myocardium after myocardial infarction. *J Clin Invest* 115: 326–338.

CHAPTER 12

Pluripotent Stem Cells

We have entered the cell, the Mansion of our birth, and started the inventory of our acquired wealth.

Albert Claude

OVERVIEW

Propensity to divide is an obvious difference between pluripotent stem cells and adult stem cells. Pluripotent stem cells are programmed to divide relentlessly; adult stem cells are programmed to divide only when called upon for replacement. The molecular mechanisms behind this fundamental difference in cellular programming are not understood but are fundamentally important to the use of both pluripotent and adult stem cells for cell replacement therapies.

The overarching consideration for any type of stem cell therapy is how to get them from the laboratory to clinical treatments. Much needs to be done. This chapter focuses on three types of pluripotent stem cells: embryonic stem (ES) cells, parthenote stem (PS) cells, and nuclear transplant stem (NTS) cells (ovasomes).

Human Embryonic Stem Cells

As of this writing, human ES (hES) cells are without a doubt the most pluripotent of all human stem cells and for that reason are the most valuable models for expansion and differentiation in the laboratory. It is important to understand hES cell biology and to also distinguish hES cells and their therapeutic potential from other potentially important pluripotent human stem cells. As described in Chapter 8, mouse ES cells paved the way for isolation of hES cells (see sidebar), and some important similarities and differences between mouse ES cells and hES cells have emerged, as listed in Table 12.1. The importance

Human Embryonic Stem Cells

Immortal, pluripotential, karyotypically normal ES cells were isolated from mouse blastocysts more than 20 years ago. Most of the scientific community viewed them as a vehicle for genetic manipulations of mice, but a small group of scientists, mostly reproductive biologists, fully appreciated the scientific and therapeutic power of that breakthrough. One of those scientists was James A. Thomson at the University of Wisconsin. Another was Alan Trounson, a pioneer in human in vitro fertilization and one of the founders of a premiere reproductive science center at Monash University in Melbourne, Australia.

James Thomson studied primate reproduction and had funding from the federal government to study pES cells. Once he had successfully isolated and partially characterized primate ES (pES) cell lines from blastocyst inner cell masses (ICMs), he knew similar work needed to be done to derive hES cells. Confusion in federal funding agencies made it impossible to proceed in a scientifically logical manner to decide how best to derive human ES (hES) cells. Instead, Dr. Thomson decided to enter into proprietary agreements with Geron, a publicly held for-profit company. Using private capital, he equipped an entirely separate laboratory "across campus" from his federally funded laboratory to avoid expenditure of federal research funds on the derivation of hES cells.

Alan Trounson pioneered many human reproduction methodologies and began dreaming of isolating stem cells from human blastocysts at least a decade ago. His research efforts in Australia were also thwarted by strict government regulations about research with human embryos. He teamed with scientists in Singapore and Jerusalem to conduct his initial attempts.

Perhaps because "across campus" was logistically easier than the Australia-Singapore-Jerusalem connection, James Thompson was the first to report successful isolation and continuous culture of hES cells in 1998. He not only characterized their chromosomal makeup, but also described the types of tissues that comprised the teratomas that resulted from transplantation of hES cells to appropriate mouse hosts. Importantly, hES cell lines appeared relatively easy to derive: from 14 inner cell mass cells, he derived 5 ES cell lines. Even before his work was reported in *Science*, the National Institutes of Health was seeking legal counsel on whether or not further research on the hES cell lines could be funded with federal dollars. The outcome of that debate would take almost 3 years. President George Bush decided in 2001 that federal dollars could be used to study "existing" hES cell lines but that no federal dollars could be used to derive more hES cell lines from human blastocysts.

Alan Trounson and collaborators published their successful isolation of hES cells from the inner cell mass of blastocysts in 2000. The characteristics of seemingly endless cycles of cell division and the capacity to differentiate into a wide variety of cell types were in agreement with the Wisconsin cell lines. Although this area of biomedical research remains highly controversial, the door to regenerative medicine has been opened.

TABLE 12.1   Comparison of Human and Mouse Embryonic Stem Cells

Characteristic	Human Embryonic Stem Cells from Eggs Fertilized by Sperm	Mouse Embryonic Stem Cells from Eggs Fertilized by Sperm
Express stage-specific embryonic antigens (SSEA)-1, -3, -4	SSEA-3, -4	SSEA-1
Express tumor rejection antigen (TRA)-1	Yes	No
Express Oct-4	Yes	Yes
Express alkaline phosphatase	Yes	Yes
Express telomerase	Yes	Yes
Factors that aid in self-renewal	Feeder cell layer plus either serum or bFGF	LIF and other factors acting through gp 130 receptor
Growth characteristics in culture	Form flat loose aggregates	Form tight rounded multilayer clumps
Can form embryoid bodies	Yes	Yes
Grow into teratomas in nude mice	Yes	Yes
Can contribute to chimeras	Unknown	Yes

bFGF, basic fibroblast growth factor; LIF, leukemia inhibitory factor.

of each of these characteristics to both of the underlying properties of stem cells, immortality and pluripotential for differentiation, is an active area of research in several laboratories throughout the world. For a variety of reasons, none of the hES cell lines currently available may prove useful for therapeutic purposes, but studies of their characteristics are invaluable guides to hES cells destined for therapeutic purposes.

The work with hES cells is just beginning, and results to date are very promising. First, although there are major differences among the various hES cells available with respect to culture characteristics and rate of self-renewal, they all appear capable of proliferation for years and hundreds of population doublings without becoming aneuploid. It is an extremely important characteristic that their chromosomal composition remains stable throughout many cell cycles.

In addition, work in laboratories in the United States, Israel, and Australia have all demonstrated that hES cells are capable of differentiating into cell types derived from all three embryonic cell layers: ectoderm, mesoderm, and endoderm. Studies demonstrating that this is also true in an animal setting, as well as a laboratory culture setting, involve injecting hES cells into mice with no immune system. Under these circumstances, the hES cells give rise to teratomas. Analyses of the cell types contained within the teratomas have revealed structures resembling gut epithelium (endodermal in origin), smooth and striated muscle cells (mesodermal in origin), and neural and squamous epithelium (ectodermal in origin) (Figure 12.1).

FIGURE 12.1.
Pluripotent stem cells can differentiate into cells characteristic of embryonic tissues. Top row, left to right: colonies of human ES cells, teratoma from primate parthenote stem cells growing in a mouse (courtesy Dr. Jose Cibelli), and teratoma removed from the mouse (courtesy Dr. Jose Cibelli). Bottom row, sections of a teratoma produced by human ES cells (courtesy Dr. M. William Lensch). Left to right, cells characteristic of ectoderm (epithelium), mesoderm (smooth muscle cells), and endoderm (gut epithelial cells).

For obvious ethical reasons, some of the experiments conducted with single mouse multipotent adult progenitor cells (described in Chapter 11) cannot be repeated with human cells. A single hES cell cannot be microinjected into a human blastocyst before transfer to the uterus to discover which tissues of the resulting baby have become chimeric for the injected cell. Because this has already been demonstrated in animal models, however, it is more important to determine whether the specific disease being addressed with stem cell therapy is ameliorated after treatment with stem cells cultured and differentiated for that specific purpose. Thus the fact that a given stem cell line can contribute to a chimera is as fundamentally important to prove for hES cells as for animal ES cells.

As described in Chapters 13 and 14, hES cells used for therapies need to be free of pathogens, tissue compatible with the recipient, capable of continual renewal in culture, and well controlled in the tissue of the donor so as to not spontaneously form tumors. These considerations support the emerging view that hES cells derived from eggs fertilized by sperm or from fetal tissues, including fetal germ cells, are not the ideal hES cells for therapeutic purposes.

Human Parthenote Stem Cells

hPS: human parthenote stem cells.

Two other types of hES cells with fewer ethical and tissue compatibility problems may hold promise for therapeutic use: those obtained from parthenogenically activated eggs (described in Chapters 4 and 5), termed **human PS (hPS)** *cells*, and those obtained from egg cytoplasts after nuclear transplantation, termed *human NTS* (hNTS) *cells*. The fundamental properties of each of these cell types have been described in preceding chapters. Development of either of these types of human stem cells has not been reported as of this writing.

Primate PS (pPS) cells have been reported, however, and results to date are highly promising. A multiinstitutional research team that began with monkey egg activation by Jose Cibelli led to the development of four parthenogenically derived blastocysts from 23 mature monkey eggs. From the four blastocysts, one pPS cell line was derived. The cells displayed the same characteristics as other ES cell lines: chromosomally normal; appear to be capable of endless cycles of growth and cell division; contain significant amounts of the enzyme telomerase, which is thought to play a fundamentally important role in maintaining the integrity of chromosomes through cell division; and display antigens thought to be associated with ES cells. The pPS cells can be induced to differentiate into a variety of cell types in culture, including primitive neurons capable of secreting dopamine, the brain cell signal missing in Parkinson disease. The pluripotent nature of the pPS cells was further demonstrated in the teratomas formed after inoculation into appropriate mouse models. Tissues from all embryonic germ layers were formed, including blood vessels, primitive gut, smooth and striated muscle cells, bone, and cartilage.

These results suggest that although parthenogenically activated primate eggs are not capable of supporting full development to an offspring, they may be capable of giving rise to pluripotent stem cell lines. In theory, hPS cells could be derived from the eggs of ovulating women with diseases amenable to stem cell therapy, such as type 1 diabetes. At the time this book was being written, there were several unknowns with this approach.

First, although there is a high likelihood that hPS cells would be antigenically identical to and thus compatible with the egg donor (syngeneic), this must be proven because of the chromosomal crossing over events that occur during oocyte maturation. It is possible that the gene rearrangements could result in expression of proteins not normally expressed by the cells of the egg donor. Even if this proves to be the case, it is likely that the antigenic differences between the cells of an egg donor and hPS cells derived from her eggs will be far fewer than antigenic differences with hES cells from an embryo. Moreover, because each egg is likely to be unique, more than one hPS cell line could be screened for therapeutic suitability. Although the technology for deriving stem cells requires development of a blastocyst with an inner cell mass, it is theoretically possible that stem cell lines in the future will be derived from individual blastomeres, which themselves could be screened for antigenic similarity to the egg donor.

Second, a homologous feeder cell layer from the egg donor needs to be developed to avoid cross-contamination of the hPS cells with potential mouse pathogens. This may or may not prove to be a significant problem. It is possible that skin fibroblasts, or cells from the ovary, may provide such a feeder layer.

The therapeutic potential of hPS cells is just beginning to be explored. This approach avoids not only all the ethical considerations associated with hES cells derived embryos or fetal stem cells derived from fetuses but may also eliminate the tissue rejection problems associated with allogeneic tissue transplantation.

Human Nuclear Transplantation Stem Cells

Analogous to animal cloning technologies, described in Chapter 10, the technical difficulties in deriving stem cells from human egg cytoplasts into which nuclei from a donor has been transplanted will undoubtedly be solved in the near future. In 2004 and again in 2005, a research team in South Korea reported the successful derivation of a total of 12 stem cell lines after nuclear transplantation of cultured skin cells from a variety of patients, including children. Unfortunately, the reports were untrue: No hNTS cell lines had been created. As of this writing, hNTS cells have not been described, although the work is ongoing in a number of laboratories.

One of the ethical stumbling blocks to federal funding for this research is the need to develop blastocysts with inner cell masses from which to isolate stem cell lines. As described previously, if the laboratory conditions for deriving stem cell lines from cleaving blastomeres could be developed, not only would the process become more efficient but the concerns of those in fear of cloning a human could be alleviated because a blastocyst would not develop. Because fertilization is not involved and the goal is not to produce an embryo but to produce a therapeutic somatic cell, a new term for this process is needed. One suggestion is **ovasome** to describe the creation of somatic cells (-some) from eggs (ova-). Accuracy of language is especially important to this new field of biomedicine. Whether or not an ovasome has the potential to develop into an offspring and therefore should be accorded the same status as an embryo is debatable, but a new term clearly distinguishes the product of nuclear transplantation from the embryo created by fertilization.

The Process of Stem Cell Differentiation

To develop into fully functioning differentiated cells suitable for therapeutic purposes, pluripotent stem cells may need to go through the same sequence of cell commitment stages as observed for stem cells in adult tissues. For example, blood stem cells first become committed to either lymphoid cell types or myeloid cell types (see Figure 1.1). As described in Chapter 1, these are referred to as progenitor cells rather than stem cells because it appears that once they have made the commitment, they do not revert to stem cells. In fact, the commitment process may be essential to preparing them for the next stage of differentiation. In addition, they may undergo a few cycles of cell division as progenitor cells. This may be an important mechanism for repopulating lost blood cells, especially in a crisis, such as hemorrhage. Studies of blood cell lineages from bone marrow and cord blood stem cells are providing important information about the number and types of cell differentiation steps that are involved in the production of mature circulating blood cells from bone marrow stem cells. This information may provide clues about the essential steps involved in cell commitment and differentiation in other tissue types.

These considerations are reminiscent of the equivalence groups described in Chapter 8. Normal tissue and organ function depends on populations of cells that can respond with precision to specific signals, both from neighboring cells and from the body at large. Such responses are generally thought to be restricted to fully differentiated adult cells, not immature progenitor cells. Given the utmost importance of balance between dividing cells and fully differentiated functioning cells that do not divide, it seems highly likely that tissues other than blood also maintain pools of cells at defined stages of commitment, ready to respond when needed. Such pools of cells could be analogous to equivalence groups, positioned to respond in one way if called upon or in another way if not called upon. The decisions could be time dependent. The nature of the response could be dependent on prior commitment events. Thus, for a mature lymphocyte to respond in a specific way, it may have had to progress through a defined pathway of differentiation events. An alternate, or deviant, pathway may result in a different response to the same stimulus.

Similar considerations may be relevant to stem cell differentiation for therapeutic purposes. For example, stem cell–derived neurons used to replace the brain cells that die in Parkinson disease (described in Chapter 13) may need to proceed through a different series of commitment and differentiation steps than stem cell–derived neurons used to replace a damaged spinal cord. Understanding the specific sequence of steps is the challenge facing stem cell therapists.

Asymmetric Cell Division

The other challenge facing stem cells is preventing both daughter cells from entering a commitment pathway; one must remain a stem cell. How this is accomplished is not fully understood, but there is growing evidence that it may relate to **asymmetric cell division**. This is a process whereby chromosomes are not randomly reallocated to daughter cells. Instead, chromosomes with "original" or "immortal" DNA strands remain with the stem cell, and the newly replicated chromosomes are allocated to the cell entering a commitment pathway (Figure 12.2).

This provocative area of study is receiving increased attention. A particularly active investigator is James Sherley from the Massachusetts Institute of Technology. He has developed mouse strains with cell markers that provide convenient probes to detect asymmetric cell division. He has pursued the concept that a well-studied DNA-binding protein, p53, plays a role in asymmetric cell division. p53 has long been associated with suppression of cancer, although the mechanism is not known. There is a connection between p53 and DNA synthesis, although the mechanism of action is not known with certainty. One enzyme that is regulated by p53 binding to DNA catalyzes the synthesis of guanosine monophosphate. Guanosine monophosphate is a component of DNA, and it also has other roles in the cell. Limitations in the supply of guanosine monophosphate prevent the cell from dividing.

A more well-known member of the p53 pathway is p21, an important member of the cell cycle pathway. The function of p21 has been studied by the knockout mouse technique (see Chapter 7). Importantly, despite the well-described obvious role p21 plays in the cell cycle, it is not an essential gene. Mice with no gene encoding p21 develop normally in utero but have few, if any, quiescent stem cells in their bone marrow. Interestingly, the lack of quiescent stem cells does not lead to total failure of blood cell production but to a higher incidence of bone marrow cell cancers. This supports the concept that limiting cell division in the stem cell population also limits

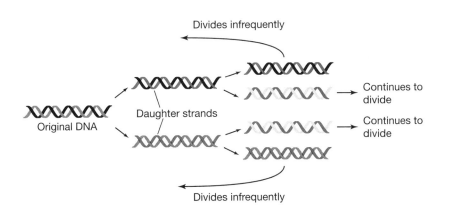

Divides infrequently

Original DNA

Daughter strands

Continues to divide

Continues to divide

Divides infrequently

FIGURE 12.2.
Asymmetric cell division.

the frequency of random mutations that give rise to the uncontrolled cell division characteristic of cancer. It follows that the daughter cells of the stem cell undergo more expansion than the stem cell population per se.

In contrast, knocking out p63, a member of the *p53* gene family, leads to neonatal death in mice because of major skin defects. This is consistent with the lack of an organized reservoir of skin stem cells that systematically replace sloughing skin cells. Knocking out another gene implicated in adult stem cell kinetics, *Pten,* results in earlier death; mouse embryos develop only about half-way. Importantly, their death appears to result from over-proliferation of brain cells, with evidence of loss of control of programmed cell death.

These considerations highlight the complexity of embryonic stem cell dynamics. There is an emerging notion that as tissues become organized, they reserve a stem cell population in keeping with their recurring daily needs for new cells. That stem cell population may, by definition, be the reservoir of DNA strands sequestered by asymmetric cell division early in development. Importantly, this may be the cell that is reprogrammable after transplant into an egg. If true, the efficiency of nuclear remodeling of any specific cell type may be directly related to the frequency of immortal DNA-containing stem cells in that tissue.

Gene Expression in Differentiating Embryonic Stem Cells

Unlike adult tissue stem cells, ES cells appear to undergo continual division in the laboratory. Perhaps cultured hES cells have lost the controls for asymmetric cell division or perhaps there is a mixed population of cells: those with immortal DNA and those daughter cells without immortal DNA that continually divide. In either case, therapeutic considerations may be independent of immortal DNA because the goal is to develop fully functioning, nondividing, adult cells.

Learning to control differentiation pathways of hES cells in laboratory culture conditions is fundamental to the goals of regenerative medicine. Understanding the sequence of gene expression—what genes are expressed when—appears to be essential to successful, sustained cell differentiation.

Several approaches have now been developed to detect and identify all the genes expressed in a given cell. Studies comparing mouse and human ES cells before and after differentiation are beginning to accumulate. The results are revealing families of genes conserved between mouse and human ES cells, and families of genes that differ. As each cell lineage is studied, the minute details of changes in gene expression will become known.

A comprehensive study conducted in the Rockefeller University laboratory of Ali Brivanlou used an array of most, if not all, human genes to determine which ones were expressed before and after differentiation of hES cells. No particular cell differentiation pathway was chosen, and spontaneous differentiation was allowed to proceed for 26 days by removing the fibroblast feeder layer.

TABLE 12.2	Gene Expression in Pluripotent Relative to Differentiating hES Cells
Up-regulated	Oct3/4, TDGF-1, Id-1H, Jumonji, Lefty A, FGF2, FGF13, Sprouty, Hey2, Sox2, THY 1, ADCY2, BIRC5, CHK2, MLF1, PAK1, PTTG1, LCK, LDB2
Down-regulated	AFP, IGFBP2, TPM2
Unchanged	G3PD, Actin

Table 12.2 lists a few of the genes expressed at different levels before and after spontaneous differentiation and those unchanged. Overall, a total of 918 genes was determined to be over-expressed in pluripotent hES cells.

Importantly, 28% of up-regulated genes had an unknown function, in agreement with the large group of unidentified genes in early mouse embryos described in Chapters 5 and 6. Table 12.3 lists the percentage of broad classes of known genes that were up-regulated in pluripotent ES cells.

As gene expression for each differentiation pathway (e.g., nerve, muscle, blood cell, pancreatic cell, etc.) becomes known, it will be possible to develop

TABLE 12.3 Classes of Genes Up-Regulated in Pluripotent hES Cells	
Unknown	28%
Energy metabolism	14%
Intracellular components	10%
Cell cycle proteins	8%
Transcription factors and DNA binding proteins	7%
RNA binding and processing	5%
Protein synthesis and processing	5%
Cell surface receptors	4%
Transporters	4%
Secreted factors	3%
Cytoskeleton	3%
Protein degradation	3%
Surface antigens	2%
Chromatin regulators	2%
Cell death	1%
DNA repair	1%

laboratory conditions to reliably direct differentiation of the pluripotent hES cells. This will be essential for the therapeutic approaches described in Chapters 13 and 14.

Summary

Pluripotent human stem cells are currently derived from the inner cell mass of embryonic blastocysts. Evidence from animal studies indicates pluripotent human stem cells may also be derived from parthenote blastocysts and from nuclear transplant blastocysts in the future. Tissue compatibility problems may limit the use of hES cells for therapeutic purposes, but they are valuable model systems for studies to perfect laboratory methods for expansion and differentiation down specified cell lineages. Human PS cells may be especially valuable for treating conditions of the woman donating the eggs, such as type 1 diabetes or spinal cord injury. NTS cells are required for perfect tissue matches for men.

hES cells share some, but not all, gene expression characteristics with mouse ES cells. Importantly, a large percentage of both human and mouse genes expressed by pluripotent cells have not been identified. This is an important area in urgent need of research.

By their nature, ES cells must have cellular controls different from adult stem cells. Early embryo stem cells must divide relentlessly and probably as rapidly as possible. Fetal tissue stem cells must reliably respond to the specific needs of the developing organ, which may include substantial cell death. Adult stem cells must retain unmutated copies of chromosomes and divide only when called on. This may be accomplished, at least in part, by asymmetric cell division, which apportions immortal DNA to the adult stem cell and more newly synthesized chromosomes to the daughter cells. Understanding the gene expression programs that orchestrate these behaviors will provide valuable information relevant to the potential of each stem cell class to alleviate the diseases that currently have no treatments or cures.

Review Questions (True or False)

1. Human PS cells are currently available for research purposes.
2. A fundamental difference between hES cells and human adult stem cells is frequency of cell division.
3. Asymmetric cell division is a concept that involves nonrandom distribution of chromosomes between daughter cells.
4. A significant percentage of genes expressed specifically in undifferentiated hES cells have an unknown function.
5. The quiescence of adult stem cells may protect them from the gene mutations that cause cancer.

ADDITIONAL READINGS

Jiang, Y., Jahagirdar, B., Reinhardt, R., Schwartz, R., Keene, C., Ortiz-Gonzales, R., et al. (2002). Pluripotency of mesenchymal stem cells derived from adult marrow. *Nature* 418: 41–49.

Kirschstein, R., and Skirboll, L. R., eds. (2000). Stem Cells: Scientific Progress and Future Research Directions. www.nih.gov

Marshak, D. R., Gardner, R. L., and Gottlieb, D. eds. (2001). *Stem Cell Biology* New York: Cold Spring Harbor Laboratory Press.

Niwa, H., Miyazaki, J., and Smith, A. G. (2002). Quantitative expression of Oct-3/4 defines differentiation, dedifferentiation or self-renewal of ES cells. *Nat Genet* 24: 328–330.

Sato, N., Sanjuan, I. M., Heke, M., Uchida, M., Naef, F., and Brivanlou, A. H. (2003). Molecular signature of human embryonic stem cells and its comparison with the mouse. *Dev Biol* 260: 404–413.

Sherley, J. L. (2002). Asymmetric cell kinetics genes: the key to expansion of adult stem cells in culture. *Stem Cells* 20: 561–572.

PART

V

Stem Cell Therapies

CHAPTER 13

Neurogenerative Diseases

The age-old idea that you can't make new neurons is being displaced by the notion that actually you can—if you know how to do it.

Ronald McKay

OVERVIEW

Until the 1990s, one of the central dogmas of neuroscience was that adult mammalian nerve cells could not regenerate. This was in contrast to amphibians and other lower animals that were known to regenerate nerves and even entire limbs. Not only could adult mammals not regenerate most nerve cells, they also could not repair damaged nerves.

Compelling evidence against this dogma was reported in the late 1990s by Elizabeth Gould at Princeton University. Dr. Gould took advantage of a relatively new technique for identifying cells in S phase by injecting animals with a modified DNA precursor molecule (termed BrdU) that mimics thymidine and is incorporated in its place during DNA replication. The methyl group on the thymidine was replaced with bromine, which could be identified by sensitive immunologic techniques using antibodies developed specifically against bromine incorporated into DNA. The method also detects DNA being repaired, but the difference between S phase and DNA repair can usually be distinguished by the amount of bromine incorporated. In a series of experiments reported in the late 1990s, Dr. Gould demonstrated evidence of DNA synthesis in brain slices of rodents and primates.

In retrospect, two prior reports had also challenged the dogma that nerve cells cannot regenerate. In 1965, Joseph Altman and Gopal Das, working at the Massachusetts Institute of Technology, published the results of studies that revealed fresh nerve growth in the adult rat hippocampus, but the reports were widely regarded as unusual exceptions to the dogma. However, in 1984 Evan Snyder, working at Children's Hospital in Boston, discovered that cells transfected with a reporter gene that readily stains blue had differentiated into neurons after transfer into a mouse. The neurons, which stained blue, had actually become integrated into brain tissue.

When Elizabeth Gould reported her results, the accumulated evidence convinced Fred Gage of the Salk Institute that it was time to look in human brains for DNA synthesis in neurons. Although a seminally important question, there seemed no way to inject humans with the bromine-substituted DNA precursor to repeat the Gould experiments. Amazingly, Dr. Gage learned that a nearby cancer study included injecting terminally ill cancer patients with BrdU to evaluate tumor growth. He was able to obtain permission to study the brains of the cancer patients when they died. Brain sections revealed DNA synthesis in cells in the hippocampus (Figure 13.1), which was in agreement with Dr. Gould's rodent and primate findings.

Does this mean brain stem cells arise in the hippocampus? There is little information about the origin of brain stem cells, but work reported last year by

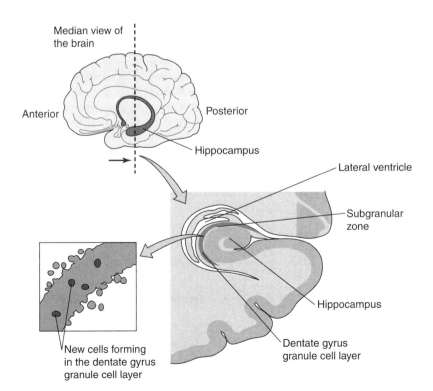

FIGURE 13.1
DNA synthesis in the hippocampus of the human brain, indicating the generation of new neurons.

Magdalena Gotz of the Max-Planck Institute of Neurobiology indicated that Pax6, a transcription factor expressed by neurons, could induce two supporting brain cells, radial glial cells and astrocytes, to differentiate into neurons. This is an important addition to the growing evidence that neural stem cells do exist. The challenge at hand is to learn to stimulate them to multiply and differentiate when needed to repair damaged nerve pathways.

In analogy to bone marrow cells, it is possible a "brain marrow" exists. This concept is supported by the growing body of evidence arising from work with embryonic stem (ES) cells. In the presence of both **epidermal growth factor** and **fibroblast growth factor**, ES cells can produce pluripotent **neurospheres**, which are aggregates of neural stem cells. Importantly, if epidermal growth factor and fibroblast growth factor are removed, the neurospheres develop into cells with many of the characteristics of astrocytes. In contrast, if the cultures are supplemented with nerve growth factor, cells with the characteristics of neurons and glial cells develop. Thus, the plasticity of nerve cells is beginning to be revealed.

Although neural fates are well mapped in lower animals, such as fruit flies and *Caenorhabditis elegans*, the sheer number of nerve cells and cell types in higher animals is staggering. Brain marrow is not as hectic as bone marrow, which pumps out tens of millions of new blood cells every second, but new brain neurons are formed each day at the rate of approximately 1 per 2,000 adult cells. It is clear that until much more is known about brain function, such as memory, it is not possible to speculate how brain cell turnover could be accomplished. How, for example, could new neurons fit into the scheme of things? Are misplaced car keys the result of a new neuron learning the ropes?

Taken together, data gathered for the past decade show that the brain makes new cells when it learns, and it makes new cells to repair damage. So, the idea of injecting cells into a damaged brain to speed up the process no longer seems as strange as it did a decade ago.

Parkinson's Disease

> At the moment the only source of unlimited numbers of the neurons appropriate for Parkinson's patients is ES cells.
>
> *Ron McKay, 2000*, Nature 406: 361–364

Parkinson's disease results from degeneration of dopamine-producing nerve cells in the substantia nigra (Latin for "black substance"), a tiny area of the midbrain that helps to control movement (Figure 13.2). Once these nerve cells stop

FIGURE 13.2
The substantia nigra in the mid-brain is the dopamine-producing area affected by Parkinson's disease.

Posterior

Anterior

Anterior

Horizontal section through cerebrum

Caudate nucleus

Corpus striatum

Putamen

Lentiform nucleus

Globus pallidus

Thalamus

Substantia nigra

Horizontal section through brainstem

Posterior

Lateral ventricles

Caudate nucleus

Corpus striatum

Putamen

Nigro-striatal neuron

Substantia nigra

Pons

Medulla oblongata

Dopamine: the neurotransmitter deficient in Parkinson's disease.

functioning, they apparently do not grow, recover, or grow back, at least not at a rate that alleviates the disease. The cells in the substantia nigra release a chemical, **dopamine**, which is one of a family of **neurotransmitters**, molecules that elicit a response from nerve cells. Substantia nigra neurons communicate with nerve cells in the **striatum**, which play a major role in controlling body movements. The result of loss of substantia nigra neurons is loss of muscle control, tremors, and stiffness. The disease usually affects people over 50 years of age, but it occasionally affects younger people as well, such as the actor Michael J. Fox. Former Attorney General Janet Reno, and the athlete, Muhammad Ali, also suffer from Parkinson's disease.

The cause of the defect in that specific type of brain cell is not known. Current treatment is focused on the administration of a chemical, **L-dopa**, which is converted in the brain to dopamine and helps to make up for the loss of dopamine production by substantia nigra neurons. A principal problem with L-dopa administration is regulating the dose to meet the need. There is sometimes too much available, which causes involuntary muscle twitching and repetitive muscle movements, termed **dyskinesia**, and at other times too little available, which results in difficulty moving at all. Moreover, after a time, the body no longer responds to the L-dopa and disease symptoms worsen. In some forms of Parkinson's disease, memory becomes impaired and patients may become demented.

Replacement therapies with neuronal stem cells could be ideal for Parkinson's disease, because there is a specific location in the brain with known types of cells needing replacement. However, to be effective, the new neurons would need to also connect appropriately with the striatum. Unless the replacement neurons can establish themselves and integrate anatomically into the brain, the transplant would be unlikely to work.

Studies transplanting adult nerve cells have never been shown to reestablish the appropriate connections. In the 1980s, however, researchers in Sweden studied the effects of transplanting fetal neurons from 7- to 9-week-old human fetuses into the brains of victims of Parkinson's disease. Similar studies had previously been performed in mice and monkeys with some improvement in symptoms. Importantly, some alleviation of symptoms was observed in some of the patients, and autopsies conducted on a few who died from causes unrelated to either Parkinson's disease or the surgery revealed survival of and outgrowth of the grafted neurons, including into the striatum. Each procedure required tissue from several fetuses, however, and the danger of tissue rejection was thought to account for some of the failures. As discussed in Chapters 1 and 11, even though these stem cells are harvested from fetuses, they are nearly completely differentiated (see Chapter 9) and may not have the same developmental potential as stem cells.

Another similar trial recently conducted in the United States demonstrated no improvement in quality of life for the patients, but autopsies of two patients who died of unrelated causes showed that many of the dopamine neurons had survived and grown. Hope remains for this type of therapeutic approach to a disease that afflicts tens of thousands of Americans annually.

Cells that secrete and respond to dopamine (**dopaminergic**) are highly specialized and quite rare in the brain. Fewer than one cell in 100,000 brain neurons communicates through dopamine release. That rarity may relate to why they are so difficult to culture in the laboratory. The limited experiments with adult brain stem cells have shown that they produce differentiated dopamine-producing neurons for a limited period of time and then stop. That has led researchers like Ron McKay at the National Institute of Neurological Disorders and Stroke, along with Lorenz Studer and Viviane Tabar at the Sloan-Kettering Cancer Center, to try ES cells. Unlike adult stem cells, ES cells can be cultured indefinitely, so they can be induced to form any number of dopamine-producing neurons desired.

As described in Chapter 8, the problem with ES cells is that they may form teratomas if they are injected while still undifferentiated. McKay and Studer have painstakingly developed procedures for culturing, differentiating, and purifying stem cell lines. They have been able to make highly purified neuronal precursors from human ES cells that do not give rise to teratomas in immune-suppressed mice. In June 2000, McKay, along with Jong-Hoon Kim and Jonathan M. Auerbach, reported on their five-stage method of inducing ES cell cultures to differentiate into neurons (Figure 13.3). Using specific chemicals, the ES cells were first induced to differentiate into pluripotential neurons. Next, two types of neurons were developed—those that produce serotonin and those that produce dopamine. After the growth factor was removed from the cells, the neurons matured into fully functional nerve cells.

In June 2002, they reported a transcription factor called nuclear receptor related-1 (Nurr1) that plays a role in the differentiation of dopamine neurons. They reasoned that if they could cause that transcription factor to be expressed in the neurons developed from ES cells, they could enrich the population of precursors for dopamine neurons. They decided to use the technique of **transfection** to ensure expression of Nurr1 by some of the ES cells.

As described in Chapters 8 and 11, constructs of DNA can be taken up by cells in culture and become integrated into the cell's chromosomes. This process is termed **transfection**. The sidebar in Chapter 8 described the use of modified gene constructs whose ends were homologous to DNA sequences at either end of a gene sequence, thus allowing for targeted integration of new gene sequences, which replace the old gene sequences in the same location. Another approach, described in Chapter 11, takes advantage of other types of DNA constructs that include at either end the DNA sequences in some types of viruses (e.g., **retroviruses**) that facilitate integration into DNA but not at specific gene sites. Thus the integration of such constructs is more frequent than homologous recombination. The individual ES cells that take up the retroviral DNA constructs and integrate them into their genome are termed **transgenic cell lines** and are analogous to the transgenic mice described in the sidebar in Chapter 5. Such constructs can also contain transcription promoter sequences that can be manipulated by the researchers, and thus the expression of the gene can be controlled by experimental conditions rather than relying on the transcription factors produced within

FIGURE 13.3
The five-stage method developed by Ron McKay, Jong-Hoon Kim, and Jonathan Auerbach for inducing neuronal development from human embryonic stem cells.

Step 1: Clonal expansion in growth medium.

Step 2: Replate cells on slick surface to promote aggregation. Spherical embryoid bodies form. Grow for four days.

Replate into original growth medium for 24 hours.

Step 3: Replace growth medium with ITSFn. Only those cells expressing nestin will survive. Grow for 10 days.

Step 4: Replate the nestin positive cells and add bFGF, laminin, SHH, and FGF-8.

Step 5: Remove bFGF from the growth medium, causing cells to stop dividing.Grow for 16 days to create dopamine neurons…

…or remove FGF-8 from the growth medium instead of bFGF to create sertonin neurons.

the cell itself. For example, a well-characterized transcription promoter sequence responds to the presence of steroid hormones. Thus adding the hormone to the culture medium leads to expression of the newly inserted gene. Because ES cells have the capacity to divide a seemingly endless number of times, transgenic ES cells can be individually selected and separated out into individual cultures. This process is termed **clonal expansion** (see Chapter 11). McKay and coworkers selected ES cell clones transgenic for Nurr1 and subjected them to the five sequential stages of culture conditions that had previously been shown to generate dopamine-producing neurons (Figure 13.3). It worked.

They needed to test their cell lines in an animal model. Rats that have been chemically treated to kill dopamine neurons provide a good model for Parkinson's disease. After this treatment, the rats have impaired motion, analogous to a human Parkinson's disease patient. Using these rats, they grafted approximately half a million of the new neurons transgenic for Nurr1 into the rat's midbrain. The hope was that the mouse neurons would appropriately populate the human brain and not form a teratoma.

Amazingly, the mouse cells were accepted into the rat's brain and thrived. They displayed the electrical properties expected of midbrain neurons. They grew into the brain and out of the original area of the graft. McKay's team could even show that new synapses were being laid down and that the mouse cells were "wired in" to the rat circuits. Most impressive was the rats' behavior. Rats with only half of their brain chemically lesioned favor the paw opposite the lesion. That was exactly what they saw in the control group. However, in the grafted rats, they noticed a remarkable improvement, close to what would be expected of normal nonlesioned rats. Thus, based on behavioral tests, the rats were shown to have accepted the mouse cells and were using them to offset the damage of their chemically induced Parkinson's disease.

The rats were examined for tumors and none was found, implying that the transplanted cells were free of undifferentiated ES cells. This experiment is vivid proof of the principle that ES cells can provide an unlimited source of controlled cells that can integrate into a host without causing tumors. To avoid tissue rejection, the ideal therapy would use stem cells derived from the patients themselves, such as through nuclear transplantation, described in Chapters 7 and 10.

Another approach to treating brain diseases such as Parkinson's disease is to deliver appropriate drugs to the site of the injury. The blood–brain barrier acts to prevent many large molecules from reaching the brain. But in December 2005, a study by Clive Svendsen and Soshana Behrstock at the University of Wisconsin showed that engineered brain stem cells can effectively integrate into the brain and express growth factors with a large molecular weight. The drug they chose was glial cell line–derived neurotrophic factor (GDNF), which has shown promise in Parkinson's disease patients but is unable to cross the blood–brain barrier.

The researchers genetically engineered human fetal brain cells to express extra GDNF and then cultured them. When these cells were engrafted into the striatum of rats and primates, they survived and secreted GDNF continuously for up to 3 months. The GDNF worked as hoped, supporting the recovery and growth of new nerve fibers. Svendsen plans further research to gain better control over the expression of these proteins. Ideally, a smaller drug that can pass through the blood–brain barrier could be used to turn on or off the expression of growth factors.

The Wisconsin researchers used fetal cells, and because they were relatively mature human cells, they were larger than the rat-host counterparts. Similar or better results might be expected for ES cells, which could be cultured to higher numbers before being coaxed to differentiate into brain stem cells. To test this idea, Fred Gage and his colleagues at the Salk Institute for Biological Sciences set out to transplant human ES cells into other animals. In December 2005 they reported that they had successfully injected human ES cells into the brains of fetal mice while they were still in the womb. These cells went on to differentiate into interconnected brain cells that, although human, had responded to the mouse environment by limiting their growth.

Of 100,000 cells injected, only a small number actually took root, amounting to less than 1% of the mouse brain cell population. There were no behavioral anomalies in these otherwise ordinary mice. Research animals with small numbers of implanted human neurons promise to offer a valuable test bed for experiments and drug testing.

Moving to monkey models of Parkinson's disease, a team of Japanese researchers headed by Jun Takahashi reported in 2005 that monkey ES cells can be induced to form large numbers of dopaminergic neurons using the fibroblast growth factors FGF20 and FGF2. When these prepared cells were implanted into monkeys with a chemical model of Parkinson's disease, they found a noticeable attenuation of neurologic symptoms.

Due to their ability to culture large numbers of cells with great plasticity, the authors of this study assert that ES cells represent the most promising avenue for Parkinson's disease research. However, commenting on this study, J. William Langston from the Parkinson's Institute in California noted that only 1% to 3% of the transplanted ES cells survived, lower than the numbers seen in human experiments with fetal cells. More research needs to be done to see if the difference is due to the source of the cells or the difference between monkeys and humans.

Alzheimer's Disease

There is no treatment for Alzheimer's to reverse the disease or restore function once it is lost. . .

Ronald Petersen, M.D., Mayo Clinic, Rochester, Minnesota

Alzheimer's disease was first described by Dr. Alois Alzheimer in 1906. At autopsy, he noticed tangles and clumps called plaque in a patient with dementia. He described an ailment that started with memory loss and disorientation and led to language impairment, delusions, hallucinations, and death. The disease can take as little as 3 years or more than 20 years to run its course. Although many other diseases can cause dementia, Alzheimer's disease is by far the most common cause. Over a third of all people over age 85 have Alzheimer's disease. As that population group increases in number, Alzheimer's disease is expected to become a major medical issue around the world. Four million Americans have Alzheimer's disease today. That number is expected to triple in the next four decades.

The hippocampus, where memories are laid down, and the frontal lobes, the seat of logical thought, are affected first. Unlike Parkinson's disease, which affects neurons that communicate using dopamine, the affected nerves in Alzheimer's disease produce a different neurotransmitter, **acetylcholine**. These cells also send out long axons into distant brain structures. Early disease may be fairly localized, but as it progresses Alzheimer's disease takes its toll in all areas of the brain. For stem cells to work, they need to do more than merely take up residence in one area, such as the substantia nigra, as in the case of Parkinson disease. They need to migrate to all the various structures of the brain that are damaged and integrate into the existing network of thickly connected neurons. Fortunately, this fits the characteristics of a neural stem cell. Apparently, in Alzheimer's disease patients the normal repair system is just not fast enough to compensate for the ravages of the disease. Stem cells could provide the boost needed to balance the ongoing destruction, thus alleviating symptoms.

Recent research has found a genetic component to Alzheimer's disease. A specific allele of the apolipoprotein E gene predisposes a person toward Alzheimer's disease. That means if the stem cells could be engineered to eliminate this allele, such as through the homologous recombination described in Chapter 8, they might halt the onslaught and even reverse the damage.

Unlike Parkinson's disease, there have been no fetal cell trials in humans yet, but research with mice is quite encouraging. In April 2001, Kiminobu Sugaya, at the University of Illinois in Chicago, reported injecting aged memory-impaired rats with neural stem cells. He found that the stem cells spread throughout their brains and that existing cells seemed to be rejuvenated. The rats were trained on a water maze, and the injected rats performed better than the controls. In fact, some of the elderly rats actually performed better than younger rats without memory impairment. One of the interesting aspects of this research is that human ES cells were used as a source for the transplants. This implies that ES cells may exhibit species-independent plasticity—finding a way to fit in, even in a different animal.

In 2003, a group of researchers led by Yoichiro Abe and Keisuke Kouyama at KEIO University School of Medicine, Tokyo reported on a technique to make Alzheimer's disease cells from ES cells. They genetically engineered the ES cells to express the V642I mutation that leads to Alzheimer's disease. This provides an excellent model to study Alzheimer's disease, highlighting the ability of ES cells to advance experimentation, as well as therapy. They believe their protocol is flexible enough to provide models for many neurodegenerative diseases in humans as well.

Multiple Sclerosis

"It's all about myelin repair and protecting nerve fibers. The goal is to translate bench research into clinical application."

Ian Duncan, University of Wisconsin

Besides the neurons and the astrocytes, there is a third important cell type in the brain called the **oligodendrocytes**. These cells send out processes to the nerve axons and insulate them from the surrounding cytoplasm. The insulation is a multi-layered lipid membrane called the **myelin sheath,** and without it nerve impulses quickly damp out, slowing or even halting communication (Figure 13.4). Diseases that involve a lack of myelin include **multiple sclerosis** (MS), **Krabbe disease,** and **leukodystrophy.** MS, a disease affecting 2.5 million people worldwide and almost half a million in the United States, is caused by a misguided immune attack on **myelin** in the brain and spinal cord. This punctures the long-held belief that the central nervous system is an immunologically privileged site. In fact, under certain circumstances, the **central nervous system** is as robustly policed by the immune system as the rest of the body.

Myelin: the insulating coating around neurons.

The disease is most prevalent among people from 20 to 40 years old and is two to three times more common in women than in men. The etiology is not well understood and may have a genetic component; relatives of MS patients are 10 times more likely to develop the disease than the public at large. But it may have an environmental component as well: MS is five times more common in northern latitudes than in the tropics. Other syndromes involving demyelination include Krabbe disease, **adrenoleukodystrophy, adrenomyeloneuropathy, progressive multifocal leukoencephalopathy, acute disseminated encephalomyelitis, central pontine myelinolysis,** and a type of MS confined to the spinal cord called **transverse myelitis.**

FIGURE 13.4
Oligodendrocytes wrap myelin around the axons of nerve cells to insulate them and prevent cross-talk.

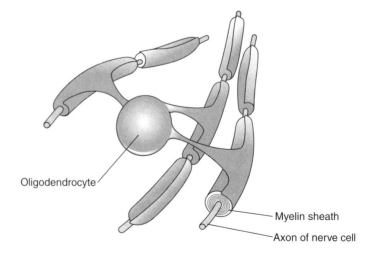

Oligodendrocyte

Myelin sheath

Axon of nerve cell

The modern view of the central nervous system immune system highlights the role of cells called **microglia**, some of which are established in the fetus and others which can circulate through compartments of interstitial fluid called the perivascular spaces. In response to injury, these cells migrate to the blood vessels and form macrophagic cells. In MS, the escalation of the immune response ultimately results in a rebuilding of the **perivascular spaces** into organized lymphoid tissue.

Late-stage MS can also lead to axonal degradation, resulting in numbness, cognitive impairment, blindness, and paralysis. Because the disease is multifocal, the question has been whether stem cells can penetrate all the spaces necessary to remyelinate axons and repair nerve damage.

In 1999, Evan Snyder reported success using fetal human stem cells to treat knockout mice that had been bred to be deficient in myelin. These mice, called shiver or twitcher mice, are born normally but develop tremors by the age of 2 weeks. Snyder injected the stem cells into the brains of newborn mice. The transplanted stem cells differentiated into oligodendrocytes and proceeded to wrap the naked axons with myelin. The cells migrated to all the parts of the brain that would be expected in normal development. In 60% of the mice, the tremors almost completely disappeared.

Snyder was able to expand his batch of human neural cells to quantities suitable for mice, but it still is not clear that these adult stem cells can be proliferated indefinitely. Nevertheless, his research indicates that given a suitable quantity of neural stem cells, some of these dreaded demyelination diseases may finally be treatable.

Ron McKay, along with Oliver Brüstle, a neuropathologist at the University of Bonn, has duplicated the Snyder research using neural stem cells derived from ES cells. McKay was looking for a plentiful supply of cells and once again found ES cells to be the most reliable source.

In 2000, a group led by John McDonald at the Washington University School of Medicine showed that consistent oligodendrocyte cultures can be produced from ES cells exposed to retinoic acid. After proving that these cells can myelinate axons in vitro, his group subsequently developed methods of culturing clumps of these cells called oligospheres. To see whether ES cells can successfully migrate and differentiate in vivo, the group then used rats with chemically demyelinated spinal cords. Large numbers of the transplanted cells both survived and successfully differentiated, primarily into mature oligodendrocytes. Cultured and labeled oligospheres were also transplanted into the spinal cords of shiverer mice, where the cells began to produce myelin and to remyelinate the host axons.

At the April 2004 presentation of The National Multiple Sclerosis Society, Dr. Tamir Ben-Hur of the Hadassah-Hebrew University Hospital and Jeff Bulte of Johns Hopkins reported on their research using human ES cells in both rats and mice. These animals were induced with experimental allergic encephalomyelitis, which mimics MS. Using magnetically tagged ES cells, they determined that the labeled cells migrated into brain lesion sites. This study introduces a useful noninvasive technique for tracking ES cell migration in the brain.

In August 2005, signaling a renewed interest in ES therapies for MS, a group led by Ian Duncan at the University of Wisconsin was awarded $3.4 million by the National Multiple Sclerosis Society to develop techniques to repair myelin damage using adult and ES cells. They are attempting to convert the preliminary bench results of these various investigators into viable therapies.

One of the problems that must be solved before any ES therapy can be made routine for MS is the immune response to allogenic cell grafts. As described in Chapter 12, this is an area that nuclear transfer stem cells can address, because they contain only DNA from the patient, hopefully avoiding immunogenicity altogether. Another important problem to resolve is the issue of teratomas. In rat experiments, it is not uncommon to observe fatal teratomas that result from direct injection of undifferentiated ES cells. As mentioned elsewhere in this chapter, the issue may be addressed by grafting only ES cells that have already been induced to form oligodendrocytes or their immediate precursors.

A final problem involves the mechanism of the disease itself. Will the immune system, already primed to attack its own oligodendrocytes, also attack any newly grafted cells? Although rats have shown that grafts can be long-lasting, the same is not known about human ES grafts. It is hoped that further studies will clarify some of these important issues.

Other Brain Syndromes

> *It is absolutely vital to continue research using embryonic neural stem cells. It may be that, for reasons we don't yet understand, adult stem cells will never be useful in therapy and that we will always need embryonic cells. Or, it may be the other way around. We just don't know.*
>
> *Charles F. Stevens*

There are many other problems with brain development or diseases of the brain that may be helped by ES cell therapy. In the United States, federal research with human embryos is severely restricted, but at the same time we are now able to save babies that have gestated for less than 23 weeks. A baby born before 30 weeks of gestation has a 60% chance of persistent neurologic problems ranging from minor learning disorders to **cerebral palsy, spina bifida, Down syndrome,** and **anencephaly.** Birth defects are the leading cause of death in the first year of life.

Prenatal ischemic brain injury is one of the precipitating factors for cerebral palsy, which is actually becoming more prevalent due to better survival rates of premature babies. ES cells may be well suited to help repopulate brains starved of oxygen and, if implanted early, may integrate well into the brains of these children and spare them the many debilitating problems associated with cerebral palsy. Animal studies have lent support to this proposition, showing a significant improvement of motor function when ES cells are implanted into fetuses. At the 34th annual meeting of the Child Neurology Society in September 2005, James E. Carroll and Cesario Borlongan at the Medical College of Georgia presented the results of a study of rats treated with stem cells. Using a model for ischemic brain injury that mimics some kinds of cerebral palsy, they injected stem cells derived from rat bone marrow directly into the brains of 22 rats. Only 1% to 2% of the transplanted cells survived, but that was enough to show a significant enhancement by day 14. Among other measures of success, the animals exhibited a 25% improvement in balance as compared with the untreated control group.

At the time of **ischemic** injury, cytokines are released, prompting stem cells to infiltrate and help repair the area. However, the cytokines only persist for a few days, possibly leading to inadequate healing. Dr. Carroll is experimenting with the use of extra cytokines along with the stem cell injections to see whether they can enhance engraftment. One direction of this research is to genetically engineer the stem cells to produce extra cytokines before injecting them into the site of the injury.

Another neurologic problem of great importance is **stroke**, which can be the result of ischemia or hemorrhage. Each year 700,000 Americans succumb to stroke, which is the leading cause of serious persistent disability in the United States. Around 80% of strokes are from ischemia, typically due to a clot. Strokes are multifocal, so treatment requires the repair of several kinds of tissue.

In July 2005, a team led by Gary Steinberg at Stanford Medical School published the results of their research with stem cells in an animal model of stroke. They injected human fetal stem cells into the brain of lesioned rats and found that large numbers of them migrated to the site of the damage and differentiated into the appropriate types of neurons. Dr. Steinberg has found that adult stem cells have short lifetimes and do not seem to migrate to the proper location. He is not using ES cells in his current research due to the difficulty of obtaining them and federal restrictions on their use in humans.

When Dr. Steinberg injected the stem cells directly into the injured tissue, very little engraftment occurred, probably due to the lack of a blood supply in the simulated stroke site. However, injected a few millimeters away, the cells survived and migrated to the area of damage, traveling up to 1.2 millimeters. Further research is planned to determine whether the engraftments result in motor recovery.

Doug Kerr and Jeffrey Rothstein of Johns Hopkins are using stem cells with mice to develop treatments for **amyotrophic lateral sclerosis**, also known as **Lou Gehrig's disease**. Their animal model uses a virus called Sindbis, which destroys the lower motor neurons of the mice, causing them to drag their hindquarters when they walk. The neuronal death caused by the virus provides a good model for amyotrophic lateral sclerosis, which is characterized by a similar death of motor neurons. In November 2000 he announced that experiments with stem cells were yielding positive results.

Rothstein and Kerr injected stem cells into the spines of the paralyzed mice and gave them time to migrate. He found that the stem cells migrated to the damaged area and that about 6% of them differentiated into neurons. Most importantly, half the mice recovered control of their hindquarters and were able to stand on all fours.

Using similar techniques, scientists hope to use stem cells to treat many other disorders of the brain. Animal trials are being conducted for Huntington's disease, Alzheimer's disease, strokes, Tay-Sachs disease, and alcohol impairment, with human trials expected soon.

One of the big issues holding up stem cell therapy is that the existing ES cell lines currently approved for experimentation have been grown on mouse fibroblast substrates. As a consequence of this cross-species contamination, none of them is appropriate for human trials. In January 2006, researchers at James Thomson's laboratory at the University of Wisconsin-Madison announced that they had succeeded in growing ES cells in a culture free of animal products. Using this protocol, they have derived two new ES cell lines. Hopefully, this will lead to new opportunities to advance ES cell therapy in humans.

Spinal Cord Injury

Never before has there been such a powerful tool, such a resource that can give so much hope. And to have it just sitting here right in front of us, ready to go while all this debate rages on, is really, really frustrating.

Christopher Reeve

Approximately two million people worldwide are living with spinal cord injuries. One of the most famous was Christopher Reeve, who was paralyzed when he was thrown from a horse in an equestrian competition. Reeve not only set up a foundation, but he was a powerful advocate for stem cell research.

The studies cited previously have demonstrated that new nerve cells can be encouraged to grow and integrate with a host brain. But many scientists were doubtful that they could reconnect a damaged spinal cord. It was also realized that more than just nerve cells were required. After damage, a cyst called a **syrinx** is formed and the nerves at the injury site become demyelinated, so oligodendrocytes are also needed to infiltrate and insulate the nerve cells.

In 1999, Jonas Frisén, Clas Johansson, and Stefan Momma and their team reported that ependymal cells responding to the site of a spinal cord injury are actually stem cells. However, instead of developing into replacement neurons, they differentiated chiefly into astrocytes, which are not functionally able to reconnect the neural pathways.

Also in 1999, John W. McDonald, at the Washington University School of Medicine in St. Louis, reported positive results using ES cells on mice whose hindquarters were paralyzed from a spinal injury (Figure 13.5). Among other things, McDonald wanted to see what success he would get by waiting for a period before injecting the ES cells. Nine days after the injury, the mice were injected with neuronal precursors that had been derived from ES cells treated with **retinoic acid.**

Retinoic acid: a molecule related to vitamin A that helps direct embryonic development.

Neurogenerative Diseases

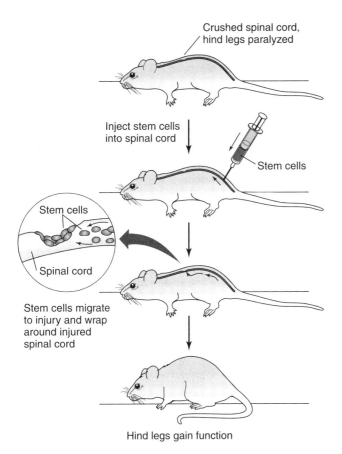

Crushed spinal cord, hind legs paralyzed

Inject stem cells into spinal cord

Stem cells

Stem cells

Spinal cord

Stem cells migrate to injury and wrap around injured spinal cord

Hind legs gain function

FIGURE 13.5
Stem cell therapy can restore function to the hind limbs of rats with spinal cord injury.

McDonald injected approximately a million cells into the syrinx that had formed around the site of the injury. The mice had been given cyclosporine to inhibit immune response and to prevent rejection of the foreign tissue.

A few weeks later, some of the mice were killed and their spinal cords were scanned for the new cells. Although many of the cells had died, several had grown projections up to 1 centimeter long. Newly sprouted axons were clearly visible. After a month, several of the mice had recovered some of their mobility and coordination. Importantly, none of the mice developed tumors.

In December 2001, Hideyuki Okano at Keio University, Japan announced that he had restored mobility to marmosets whose arms were paralyzed due to a spinal cord injury. All the marmosets, which are small monkeys, showed improvements after the transplant, and one regained almost full use of its arms. This marked the first success repairing spinal injuries in primates with stem cells. Okano used human fetal stem cells, which he was able to amplify to over a million cells, suitable for therapy in a small animal like a marmoset.

Since then, in August 2002 Okano was able to show that stem cells are more effective if they are infused a few days after the trauma. He noted that stem cells had a much better chance of differentiating into neurons if they were injected 9 days after the injury. That, coincidentally, is the same waiting period that McDonald used in his experiment. Immediately after an injury, chemicals are released that inhibit neuron growth. After 9 days, these inhibitors have dissipated and neuronal growth can proceed.

Okano was able to obtain fetal neural stem cells from aborted fetuses. His technique involves culturing and expanding the fetal cells to get the greatest utility from them. Unlike ES cells, however, neural stem cells have not been proven to provide an immortal cell line.

In May 2005, Hans Keirstead and his team at the Reeve-Irvine Research Center at UC Irvine (named in honor of Christopher Reeve) reported on experiments using human ES cells to treat injured rats. They induced the ES cells to differentiate into early-stage oligodendrocytes and then injected them into rats with spinal cord damage. One group was injected 7 days after the damage and the other group was injected after a 10-month waiting period. In both groups, the ES cells formed functional oligodendrocytes that migrated to the site of the damage. But although the earlier group showed a significant improvement in mobility, the second group did not. Apparently, even though the oligodendrocytes established themselves, they could not penetrate the scar tissue that had formed around the dying neurons.

The oligodendrocytes that were administered after only 7 days were able to reinsulate the injured neurons and to provide them with an environment conducive to healing. This research establishes the importance of timely remyelination to the repair of spinal cord injuries. It also indicates that the body's immune response to spinal cord injury can be counterproductive. Notably, in all the experiments where ES cells were induced to form precursors of nerve cells, teratomas were not observed. This indicates that ES cells are plastic enough to become multiple cells lines without devolving into tumors.

Summary

The central dogma of neuroscience has been that mammalian neuronal pathways cannot regenerate, but reports in the 1990s indicate they can, which has given hope that growth and repair in the central nervous system can be encouraged with stem cell transplants. Parkinson's disease is one of the diseases that seems suited to stem cell therapy, because it is the result of the loss of dopamine-producing neurons. Preliminary trials with fetal nerve cell implants produced encouraging results for some, but not all, recipients. Research on rats has shown that properly differentiated ES cells can reverse the effects of Parkinson's disease without forming teratomas.

Alzheimer's disease is another disease that may yield to stem cell therapy, if ES cells can be coaxed into repairing acetylcholinergic neurons faster than they are destroyed. As with cardiac failure, the natural healing mechanism seems to be there, but it is just too slow to be effective. It is also possible that Alzheimer's disease is a genetic defect, which may be amenable to recombinant ES cell therapy.

Myelin surrounds nerve cells like an insulator around an electrical wire, preventing cross-talk. Diseases that attack the oligodendrocytes that form the myelin sheath include MS, Krabbe disease, adrenoleukodystrophy, and many more. Ultimately, this lack of insulation often leads to neuronal damage, resulting in complicated multifocal diseases. ES cells, however, are pluripotent, and work to date indicates that they can differentiate into the specific cell types required to address these diseases.

Experiments with a rat model of cerebral palsy have shown significant improvement with engrafted adult stem cells. Similarly, animal models of stroke have shown that ES cells can migrate, engraft, and differentiate into appropriate tissues at the site of the damage.

Stem cells may also have a role to play in spinal cord damage. Multiple lines of research have indicated that ES cells can be differentiated into oligodendrocytes that wrap around damaged axons and help to heal the lesion.

Review Questions (True or False)

1. Brain cells do not regenerate, and therefore damage to the central nervous system is irreversible.
2. Oligodendrocytes can be induced to form from ES cells by the application of retinoic acid.
3. Parkinson's disease is a disease of old age.
4. Parkinson's disease is caused by the loss of acetylcholinergic neurons.
5. Alzheimer disease is caused by the loss of dopaminergic neurons.
6. Stem cells, following the trail of cytokines released by damaged tissue, can migrate to those sites and differentiate into cells appropriate for healing.
7. Cells called oligodendrocytes are responsible for forming the myelin sheath surrounding nerve cell axons.

ADDITIONAL READINGS

Altman, J., and Das, G. D. (1965). Autoradiographic and histological evidence of postnatal hippocampal neurogenesis in rats. *J Comp Neurol* 124: 319–333.

Eriksson, P. S., Perfilieva, E., Bjork-Eriksson, T., Alborn, A. M., Nordborg, C., Peterson, D. A., and Gage, F. H. (1998). Neurogenesis in the adult human hippocampus. *Nat Med* 4: 1313–1317.

Freed, C. R., Greene, P. E., Breeze, R. E., Tsai, W. Y., DuMouchel, W., Kao, R., Dillon, S., Winfield, H., Culver, S., Trojanowski, J. Q., Eidelberg, D., and Fahn, S. (2001). Transplantation of embryonic dopamine neurons for severe Parkinson's disease. *N Engl J Med* 344: 710–719.

Jong-Hoon, K., Auerbach, J. M., Rodríguez-Gómez, J. A., Velasco, I., Gavin, G., Lumelsky, N., Lee, S-H., Nguyen, J., Rosario Sánchez-Pernaute, R., Bankiewicz, K., and McKay, R. (2002). Dopamine neurons derived from embryonic stem cells function in an animal model of Parkinson's disease. *Nature* AOP, published online 20 June 2002; doi:10.1038/nature00900.

McDonald, J. W., Xiao-Zhong, L., Qu, Y., Su, L., Mickey, S. K., Turestsky, D., Gottlieb, D. I., and Choi, D. (1999). Transplanted embryonic stem cells survive, differentiate, and promote recovery in the injured rat spinal cord. *Nat Med* 5: 1410–1412.

Sang-Hun, L., Lumelsky, N., Studer, L., Auerbach, J. M., and McKay, R. D. (2000). Efficient generation of midbrain and hindbrain neurons from mouse embryonic stem cells. *Nat Biotechnol* 18: 675–679.

Wakayama, T., Tabar, V., Rodriguez, I., Perry, A. C. F., Studer, L., and Mombaerts, P. (2001). Differentiation of embryonic stem cell lines generated from adult somatic cells by nuclear transfer. *Science* 292: 740–743.

Tissue Systems Failures

ES cells . . . could provide a potentially unlimited source of islet stem cells that would be relatively straightforward to isolate and manipulate in culture.

Jon Odorico

OVERVIEW

Before any stem cell therapies can be considered for human trials, an important problem must be addressed. Currently, human embryonic stem (ES) cells are cultured on a thin mat of mouse fibroblast cells. To ensure that no animal products or viruses are communicated from this mat to the human ES cells, a new culturing protocol must be established.

In 2002, scientists in Singapore managed to coax human cells into providing the proper growth environment. The technique, developed by Ariff Bongso and his team at the National University of Singapore, also has the bonus of extending the culturing period from 7 to 9 days. This is the amount of time a cell culture can grow without spontaneously differentiating. To continue the line, cell cultures must be broken up and separated into different dishes. By extending the time between passages, a far greater number of cells can be cultivated between passages.

In January 2006, Tenneille Ludwig at the laboratory of James Thomson reported on their development of a defined culture system for ES cells that is free of animal cells. Using this system, they derived two new human ES cell lines. Although research progress has been made with private funding in America, the pace is slow. Countries like Saudi Arabia, Israel, and Singapore, where the regulatory environment is favorable to stem cell science, will take the lead and determine the direction of research. Because President Bush has declared that federally funded research can only be performed on an existing set of embryonic cell lines—all of them grown on mouse fibroblasts—human clinical trials are unlikely to happen in this country until human parthenote stem cells or human nuclear transplant stem cells described in Chapter 12 are developed.

Diabetes

Diabetes affects over 170 million people worldwide, approximately 25 million of whom live in the United States. It is a disease characterized by the inability to either manufacture or use insulin, the protein hormone that regulates the uptake of glucose into cells. **Insulin** is secreted by specialized cells within the pancreas that are organized into regions named the islets of Langerhans. Without insulin, undigested sugars build up in the blood plasma instead of entering the glucose pathway (described in Chapter 6) within cells. Because glucose metabolism is a principal source of energy for human cells, the cells and the tissues they comprise are chronically energy deprived. Over time this causes widespread damage to tissues and organs.

Insulin: the protein hormone responsible for glucose uptake into cells.

There are two principal types of diabetes. Type 1 is also referred to as juvenile-onset diabetes and is characterized by undetectable levels of circulating insulin due to an absence of islet cells in the pancreas. Type 2 diabetes usually affects adults and is characterized by nearly normal levels of circulating insulin that do not stimulate normal glucose uptake by the cells. Type 2 diabetes accounts for at least 90% of U.S. cases and can often be controlled by diet and exercise. One in five U.S. residents over the age of 65 has diabetes. Because the exact mechanism of type 2 is still unknown, much work still needs to be done before a cure can be developed.

Type 1 Diabetes

On the order of 1 in 200 U.S. children has type 1 diabetes. Because type 1 diabetes involves cell loss, it is a likely candidate for stem cell treatment. Stem cell therapy would involve differentiating stem cells into islet cells, or islet cell precursors, and then transplanting them into the patient. With an appropriate source and supply of islet cells, type 1 diabetes should be possible to contain and even cure.

It has been shown that type 1 diabetes can be cured with islets transplanted from donated organs in a procedure known as the Edmonton protocol. Unfortunately, the treatment calls for two pancreases per patient, and there are not nearly enough donated organs to go around. Only a few hundred of these procedures are possible each year, reaching only a small fraction of needy diabetics. Nevertheless, the success of the procedures provides proof that transplanted islets can integrate with the host tissue and function properly, actually curing this previously intractable disease. The successes immediately raised the need to isolate pancreatic stem cells and to develop the methods to differentiate them into islet precursors to increase the supply of cells for therapy.

Unfortunately, pancreatic stem cells from adult organs have been frustratingly difficult to identify and isolate. Scientists do not yet know their morphology or what cell surface markers they display. Pancreatic stem cells also appear to have a fleeting existence in the developing fetus, so researchers are turning to ES cells, both to better understand the genesis of these cells and to obtain sufficient quantities for therapy. Nonetheless, whether the islet cells arise from adult tissue stem cells, fetal tissue stem cells, or ES cells, the immune system of the recipient must be suppressed to avoid rejection of the transplanted cells. Thus, a lifetime of injecting insulin is replaced by a lifetime regimen of immune-suppression drugs. Although the side effects from immune suppression are more controllable than the tissue pathology that results from type 1 diabetes, it is clear that the ideal solution is to derive pancreatic stem cells from the diabetic's own cells.

Toward this end, several researchers have been developing protocols to induce the growth of pancreatic islet cells from ES cell cultures. In 2001, Bernat Soria, at the Universidad Miguel Hernández, was able to create insulin-secreting cells from mouse ES cells. Soria prepared a DNA construct that included a reporter gene (*lacZ*) and DNA sequences homologous to the flanking sequences of the insulin gene so that the DNA construct could be incorporated into chromosomes by homologous recombination, as described in Chapter 8. The insulin gene sequences were carefully chosen so that expression of the construct would be under the control of the promoter sequence that controls insulin production in islet cells. In this way, she hoped to easily detect islet cell precursors at the first stage that they expressed insulin. Mouse ES cells were transfected with the DNA construct. The cells were then allowed to form embryoid bodies, with the hope that insulin-producing cells would differentiate out of the mass. It worked. Cells were selected for insulin production.

Soria injected the carefully selected cells into the spleens of mice with chemically induced diabetes. Within 24 hours, the mice were producing their own insulin and normalizing their glucose levels. Four weeks later, 60% of the mice were still normalized, indicating that the stem cells had successfully integrated into the bodies of the host mice. Twelve months later, some of these mice were still controlling their glucose levels, essentially cured of diabetes.

As promising as these results were, the ES cell procedure remains highly inefficient, yielding insulin-producing cells once for every 100 attempts. Unfortunately, the cells that grow well do not produce therapeutic levels of insulin, and those that produce insulin do not grow very well. Soria notes, "There appears to be an inverse relationship between the ability to proliferate and the ability to differentiate. We have not yet solved this problem, which again shows how much we need more research." Moreover, these studies also demonstrated that isolated cells do not produce the same smooth adjustments to glucose levels that are produced by islets that include three other cell types. This further complicates the production of appropriate islet cell replacements.

In February 2001, Jon Odorico and James Thomson, of the University of Wisconsin, demonstrated that primate ES cells can spontaneously differentiate into pancreatic cells. Odorico has also done work with human ES cells, but the work is progressing slowly. Odorico isolated a batch of cells that he believes are beta cells. They have been stained for analysis, but because of federal regulations he cannot evaluate them with university microscopes. The results of his studies await private funding.

In April 2001, Ronald McKay and Nadya Lumelsky, of the National Institute of Neurological Disorders and Stroke, announced that they had been able to reliably differentiate insulin-producing cells from mouse ES cells. They allowed ES cells to form embryoid bodies and looked for those cells that expressed the protein nestin. **Nestin** is usually considered to be a neural cell inducer, but its presence is also associated with pancreatic islet cells. Using their five-stage culturing protocol (see Chapter 13), they induced the nestin-producing cells to assemble themselves into three-dimensional clusters that looked and acted like fully formed islets. When glucose was added to these structures, they produced insulin in response.

Nestin: the protein expressed by progenitors to neurons.

These islets were then transplanted into the shoulders of diabetic mice, where they were quickly vascularized and supported. Unfortunately, the cells secreted a low level of insulin and were unable to totally cure the mice of diabetes. Nevertheless, this study showed that fully formed islets, composed of all four cell types and even including neurons, can be assembled out of ES cells. Note also that the locus of cell injection may not be important. Soria transplanted into the spleen, and McKay's group simply chose the shoulder. Both locations tolerated the cells, vascularized them, and allowed them to produce insulin.

In August 2001, Suheirr Assady, at the Technion in Israel, showed that human ES cells could also be induced to form insulin-producing cells, taking another step on the path to human trials. Assady says, "These findings validate the human ES cell model system as a potential basis for enrichment of human beta cells or their precursors, as a possible future source for cell replacement therapy in diabetes."

In a 2005 review of the literature, Luc Bouwens at the Vrije Universiteit Brussel in Belgium points out that the insulin-producing cells derived to date are low in number and low in insulin production. Their studies indicate that production of islet precursors is blocked at many stages of development and that more research is needed to pin down the various transcription factors that yield high-quality pancreatic cells.

Additional studies have elaborated on the cascade of events necessary to induce islet precursors from ES cells. In January 2006, a team headed by Frédéric Lemaigre of the Institute of Cellular Pathology in Brussels, reported that a sequential activation of hepatocyte nuclear factors (vHNF1 and HNF6) and pancreatic and duodenal homeobox (Pdx) transcription factors appears to guide the induction of pancreatic precursor cells. This cascade may be useful for in vitro amplification of islet cells or as a tool to identify and separate pancreatic cells from other ES cells in culture.

Type 1 diabetes is thought to be an autoimmune disease in which islet cells are targeted for destruction by the victim's own immune system. If the diabetic's own islet cells were replenished, the risk is that the immune system might destroy them just as it did the original islets. However, it might be possible to genetically engineer these cell cultures so they will not be attacked by the immune system. If the genes that encode the cell surface proteins that target the islet cells for destruction could be excised, the islets might escape detection.

This theory was supported in 1993, when Peter G. Stock and his team at UCSF showed that islets from mice lacking the major class of cell surface proteins that usually stimulate immune rejection of transplanted tissues were not rejected. Therefore, a potential stem cell therapy might involve using a knockout vector to disable the major histocompatibility complex gene or genes. This "unmarked" ES cell line could then be cultured, cryopreserved, and stockpiled. When needed, the cells would be thawed and treated with special proteins to convert them to pancreatic stem cells. Once injected, these cells could repair the pancreas while avoiding the predations of the immune system, completely curing the disease with no side effects.

Alternatively, a therapy could be aimed at the misguided immune system itself. As in some cancer therapies, a technique may be developed to harvest a few stem cells, destroy the faulty immune system with radiation or chemotherapy, and then restart the immune system with the harvested cells. Research by Richard Burt suggests that once the immune system is renewed, it starts from scratch and no longer attacks the patient's own cells.

It may even be possible to modify autoimmunity with a simple hematopoietic stem cell transplant. Dr. Nagwa S. El-Badri and colleagues, at All Children's Hospital, University of South Florida, were able to completely halt rejection diseases in autoimmune-prone mice with a single allogenic transplant. The mice, dubbed W/B F1, typically contract one of several autoimmune diseases by 8 weeks of age. However, after the transplant, 100% of the mice survived a year, completely free of autoimmune diseases.

Type 2 Diabetes

Type 2 diabetes afflicts 10 times more people than type 1 and is a major medical problem. Most people with type 2 diabetes have insulin resistance, which is still poorly understood and may have multiple underlying causes. However, up to one-fourth of type 2 diabetics take insulin, so it may be possible to treat these cases like type 1, by injecting islet cells derived from cultured ES cells.

If, as suspected, a genetic flaw lies behind type 2 diabetes, then genetic engineering could repair the deviant gene. The corrected stem cells would be cultured and transplanted back into the patient. Because the stem cells are persistent and can spawn islet cells as needed, the diabetic might look forward to a complete cure with a single injection.

Rudolph Jaenisch and George Daley, from the Whitehead Institute for Biomedical Research, demonstrated the feasibility of this procedure using mice with a deadly genetic autoimmune defect. ES cells were created for each mouse using nuclear transfer. The flawed gene was then repaired in the laboratory. After culturing, these cells were injected back into the mouse. The transplanted stem cells were not rejected, formed stable cell populations, and partially cured the mice of their autoimmune disease.

Stem cells hold great promise for the treatment of the secondary diseases of diabetes, such as heart, kidney, eye, and circulatory ailments. As noted elsewhere in this book, each of these diseases is being targeted for research by stem cell scientists, who are hopeful for longer and healthier lives for all diabetics.

Cardiomyopathy

Cardiomyopathy, a disease of the heart characterized by weakened muscle fibers, affects over 4.8 million people in the United States alone. Most cases of cardiomyopathy are diagnosed at the time of a heart attack, which injures the heart and, due to its slow recovery time, leaves it vulnerable to further degradation.

Heart transplants are a last line of defense against a failing heart. There are far too few hearts available, and recipients must prepare for a continuing regimen of immune-suppressing drugs that leave them vulnerable to other infections.

It has long been assumed that cardiac muscle, like neural tissue, does not regenerate in adults. After a heart attack, the only observed changes seem to be cell death and the formation of scar tissue. However, new technologies have led to better observation, and it is generally now accepted that some limited regeneration does indeed occur. **Cardiomyocytes**, the cells that comprise the pumping muscle of the heart, divide and start to repopulate the area. However, the response is often too little and too late. Unfortunately, no one has identified heart stem cells, which would presumably enhance the magnitude and speed of recovery. This is supported by the work described in Chapter 11 with mouse multipotent adult progenitor cells.

Cardiomyocyte: the precursor to heart muscle cells.

Nonetheless, other researchers have reported that hematopoietic stem cells derived from bone marrow can, in vivo, differentiate into apparent cardiomyocyte precursors. In 2001, Piero Anversa showed that injecting these marrow cells into damaged rat hearts would heal the tissue. The healing included heart muscle and the vasculature that supports the heart. Both modes of healing were evidently due to the fact that the injected cells were primitive enough to differentiate into multiple tissue types.

In November 2000, Israeli researchers, headed by Nissim Benvenisty at the Hebrew University of Jerusalem, showed that human ES cells automatically generate cardiomyocyte precursors when they are allowed to form embryoid bodies. This work was extended by Lior Gepstein's group, at the Technion-Israel Institute of Technology, who showed more evidence that cardiomyocytes spontaneously appear in embryoid bodies. The scientists were careful to check expressed genes, transcription factors, enzymes, and electrochemical changes to verify that these cells were bona fide cardiomyocytes. Importantly, cardiomyocytes make themselves instantly known by their unmistakable rhythmic contractions. Without too much inducement, they simply differentiate out of mesoderm tissue and start pulsing. Gathered together by the millions and coordinated by nerve impulses, these tiny throbbing cells give rise to the impressive pumping action of the heart.

In January 2006, a team at the University of Wisconsin headed by Timothy J. Kamp reported that they had successfully injected mouse ES cells into mice with experimentally induced heart attacks. The unmodified embryonic cells differentiated into three types of heart tissue: cardiomyocytes, vascular smooth muscle and endothelial cells. Importantly, no tumors were observed. Although the hearts were not completely repaired by these three tissues, this research points the way to further studies that are needed in higher animals before human studies can be undertaken.

To repair a typical infarction, it may take a million or more cells. This argues against using adult stem cells, both because they are rare and because they are difficult to expand. ES cells are better in this regard. Large quantities of cardiac precursor cells can be generated in vitro and then injected into the injured area. Because they are multipotent precursor cells, they can quickly expand their numbers in vivo, creating all the cell types needed for proper repair. To avoid rejection, the best cells would be derived from the patients themselves, using nuclear transfer. However, there is still an important issue that must be emphasized with regard to cardiac stem cells derived from ES clones: They are needed immediately after an attack. Longer waiting times lead to less positive outcomes, but the process of producing therapeutic quantities of cardiac stem cells after nuclear transfer can take weeks. To be most useful, then, patients would have to bank their own stem cells before an infarction.

In June 2000, the American Heart Association endorsed ES cell research and considered funding some studies. The American Heart Association said that ES cell research represents "the most promising medical and scientific research" to fight cardiovascular disease. However, when they announced their intentions, a mail-in campaign spearheaded by Catholic Archbishop Justin Rigali of St. Louis convinced them to cancel their plans. As of 2002, there are still no federally funded heart studies that use human ES cells.

Kidney Failure

In June 2002, Robert Lanza, Michael West, and his colleagues at Advanced Cell Technology reported that they had created a cow kidney from stem cells produced by nuclear transfer. After transplanting the kidney tissue under the skin of the donor cow, they found that it performed certain functions of a normal kidney, including the excretion of urine.

An April 2005 study by Toshio Mochizuki at the Hokkaido University Graduate School of Medicine in Sapporo used ES cells to create a mouse model of **autosomal dominant polycystic kidney disease** (ADPKD). He and his team developed chimeric mice by aggregating PKD ES cells with mice morula. These mice were found to have kidney pathologies similar to those found in human ADPKD. They discovered that the protein **polycystin-1** may play a role in the prevention of ADPKD through its induction of p53 and JNK proteins. A surprising outcome was the finding that the normal mouse cells from the mixed morula also participated in the formation of kidney cysts, suggesting that signaling from the PKD ES cells may induce aspects of ADPKD.

An important issue addressed by this study is the role of mitochondrial DNA in tissue rejection. Because nuclear transfer uses an enucleated egg, there is no nuclear DNA from the donor in the cloned cells. However, there are mitochondria in the egg that have their own DNA and replication machinery, and they do get passed on to the cloned stem cells. Might these egg mitochondria in some way produce an immune reaction? The study showed no rejection phenomena, thereby removing an important unknown about the efficacy of therapeutic cloning.

Liver Failure

Can liver tissue be repaired by stem cells? The evidence so far is positive. In November 2000, a team headed by D. S. Krause examined a number of liver biopsies from allografted bone marrow recipients. In particular, they were looking for women who had transplants from men and vice versa. They found two female and four male candidates. By staining for the X and Y chromosome, they were able to show that donor cells from the marrow had populated the livers of all subjects. The degree of infiltration was surprising—up to 43% of cells from some livers contained opposite-sex chromosomes. The largest infiltration was in a patient who had recurrent hepatitis C. This indicates that the infiltration was motivated by the injury site itself and that bone marrow stem cells were recruited to differentiate into hepatic cells and repair the damage.

It is also known that ES cells can give rise to **hepatic** cells. Research done in 2002 by Hirobumi Teraoka showed that after allowing ES cells to form embryoid bodies, hepatic cells start to show up within 12 days. These findings were supported by a separate Japanese study by Yukio Tsunoda that also showed that ES cells could form hepatic cells. The researchers found that hepatic cells spontaneously arose in embryoid bodies at around day 14. The hepatic cells took up an indocyanine green stain and formed distinct three-dimensional shapes in the embryoid body. These cells were found to produce the families of messenger RNAs that are characteristic of liver cells. Some of the cells were injected into the hepatic portal vein of other mice, where they integrated into the host liver and formed typical hepatic cells.

A 2006 study by H. Nakauchi at the University of Tokyo further characterized hepatic stem cells. They used fluorescence-activated cell sorting and were able

Hepatic: relating to the liver.

to retrieve stem cells that can migrate into damaged liver tissue and differentiate into liver cells indistinguishable from native hepatocytes. Their techniques will help to analyze liver development in vitro and may lead to novel stem cell therapy for liver diseases

Cancer

As discussed in Chapter 1, cancer is a disease characterized by cells that no longer perform their intended function and instead become highly proliferative and ultimately mobile and metastatic. In a sense, these cells are somewhat like primitive stem cells, without a declared function. Cells become cancerous on a regular basis. In general, because cancer cells usually express proteins on their surface that are not expressed by the noncancerous cells in the tissue involved, the cancer cells are discovered by the immune system and destroyed. If the cancer cell can hide or disguise its antigens, it may escape attack. Cancer cells may also avoid attack by amplifying a population of immune suppressor cells.

In a sense then, tumor growth and metastasis may be a malfunction of the immune system. And, in fact, bone marrow transplants that are prescribed for many cancers (see Chapter 1) aim to repair the immune system. The idea is to sensitize the harvested marrow to eliminate cancer cells.

One of the earliest applications of bone marrow transplants was probably also the first to use stem cells to cure cancer. In 1956, Dr. E. Donnall Thomas transplanted donor marrow into a leukemia patient and generated a complete remission of the cancer. This pioneering research earned him (along with Joseph E. Murray) a Nobel Prize in 1990, as described in Chapter 1. Treatments similar to this initial protocol are still used today and are responsible for up to an 80% remission rate in leukemia patients.

The success of bone marrow transplants depends largely on the ability of stem cells to give rise to all the cells of the immune system. Recently, evidence has accumulated that marrow from closely, but not perfectly, matched donors may yield a better prognosis than autologous marrow (derived from the patient's own marrow). The speculation is that these cells do not have the same genetic deficiency as the patients' own cells. The problem is that the body's immune system may attack them in what is called host-versus-graft disease. Moreover, because a new immune system is being injected, there is the further problem of graft-versus-host disease, where the new immune cells target the patient.

The challenge is to find a donor graft that targets the cancer cells but not the healthy cells. Despite the knowledge base about the genetics of immune rejection, achieving the proper balance is often difficult. ES cells derived from the patient through nuclear transplantation or parthenogenetic stimulation of eggs, if the patient is a normally ovulating woman (described in Chapter 3), could eliminate a host of problems. Once ES cells have been derived from the patient, they could be amplified and induced to form blood-cell precursors, the primary regenerative component of bone marrow. Genetic deficiencies could theoretically be corrected in culture, before the cells are reinfused. In this way, patients could receive cells that are genetically identical to their own—with the exception of the corrected genes. These cells could both cure the cancer and avoid autoimmune destruction.

In October 2005, a team led by Dan Kaufman at the University of Minnesota demonstrated the feasibility of fighting cancer by boosting the immune system when they induced the generation of functional natural killer cells from human ES cells. These cells possessed the inhibitory and activating receptors found in mature natural killer cells, and when grown in culture they were able to lyse human tumor cells. This research could lead to novel cellular immune therapies.

Stem cells may have much to teach us about cancer in general and may actually be causative in certain types of cancer. In November 2005, a team led by Renee Reijo Pera and Amander Clark at UCSF reported that the expression of four ES cell genes are elevated in certain types of breast and testicular cancer. The genes they discovered in these tumors express the homeoproteins **GDF3, STELLAR, NANOG,** and **OCT4**. These genes may have a role to play in tumorigenesis, and they may also serve as predictive signals of cancer, leading to new diagnostics and therapy focus.

The UCSF team concentrated on **seminomas**, which are tumors found in 50% of testicular cancers. Seminomas are associated with sperm-producing germ cells, apparently arising early in the formation of sperm cells. What surprised the team was finding elevated expression of the same group of genes in breast tumors, which are not sperm-cell related. This team had previously reported that these same four genes are involved with early differentiation of ES cells into primordial germ cells. Seminoma cells are structurally similar to these normal progenitor germ cells but with a duplication of genes on chromosome 12. The researchers expect that these genes will be found in other nonreproductive cancers as well, especially in the light of overexpression of OCT4 in cases of intestinal dysplasia.

In January 2006, a team led by Mark Shackleton and François Vaillant at the The Walter and Eliza Hall Institute in Australia reported on their discovery of breast stem cells. They speculate that one of these stem cells could accumulate enough genetic errors to become a cancer stem cell. Although these cells are cancerous, they may replicate much slower than typical cancer cells, thereby evading the chemicals and radiation treatments designed to eradicate fast-growing tumors. They may thus serve as a repository for a subsequent attack.

These results and others are beginning to directly implicate stem cells in tumorigenesis. A view starting to develop among oncologists, called the cancer stem cell theory, posits that a small number of cells in a tumor mass are actually stem cells. Like ES cells, they divide asymmetrically, producing both a cloned daughter and a more differentiated cancer cell that constitutes the bulk of the tumor. At least two homeoproteins, OCT4 and NANOG, are responsible for maintaining the pluripotent state of normal ES cells, and these are also popping up in brain, breast, bone, and germ cell cancers. This could explain the limited effectiveness of standard chemotherapy and radiation therapy, which concentrate exclusively on rapidly dividing cells, missing the stem cells altogether.

This theory also opens the door to new cancer therapies that can block these homeoproteins or their promoters and force the aberrant stem cells to differentiate and lose their pluripotency. Greater knowledge of how these cancerous stem cells proliferate and avoid the immune system may soon lead to better treatments.

Hemophilia

Hemophilia is a genetic disease that prevents blood from clotting. Without the ability to clot, not only is death by exsanguination possible, but tissues have difficulty recovering from injury. About 17,000 Americans are afflicted by type A or type B hemophilia, which involve mutations in protein factors necessary to form a clot. Another bleeding disorder is **von Willebrand disease**, which affects up to 2% of the population and involves a dysfunction of platelets. Because the disease is sex linked, hemophiliacs are almost exclusively men, but von Willebrand disease affects both women and men. The long-standing treatment for hemophilia has been transfusions of clotting factors recovered from human blood donations. However, before the early 1980s, the blood supply was not being screened for human **immunodeficiency** virus, the causative agent of acquired immunodeficiency syndrome, and now an alarming 50% of hemophiliacs are infected by human immunodeficiency virus. Of the original 10,000 men and boys infected in the early 1980s, 7,500 have died. Any treatment that can avoid the risk of exposure to pooled human blood products would be clearly superior. Stem cell therapies have been proposed.

Because a level of even 5% of normal clotting factor can greatly ameliorate the symptoms of hemophilia, the possibility of introducing blood stem cells that give rise to clotting factors merits study. In 2000, a Stanford study showed that hemophilia in dogs can be reversed by transfection of the blood coagulation **factor IX** into skeletal muscle using a viral vector. The transecting agent, adenoassociated viral vector, has also been used in murine studies. In mice that receive injections of the factor IX–adenoassociated viral complex, the levels of factor IX are substantially increased. These studies provide evidence that hemophilia can be cured with the proper expression of a single gene.

The same group in Roslin, Scotland that produced Dolly the sheep also produced Poly, a sheep cloned from a transfected embryonic cell line. Poly produces factor IX in her milk. Although Poly was engineered to produce therapeutic quantities of factor IX in her milk, it also demonstrates that nuclear transfer can be combined with genetic manipulation to create persistent genetic changes in mammals other than cows, as described in Chapter 10.

In February 2005, a team at the University of North Carolina headed by Jeffrey A. Frelinger succeeded in grafting functioning liver cells into mice that were deficient in factor IX. Their success was apparently due to first culturing murine ES cells in the presence of fibroblast growth factor. After engraftment, these cells produced factor IX and extended the lives of these deficient mice. Interestingly, these animals survived even though the transplanted cells were from a different

Immunodeficiency: an inadequate immune response to pathogens.

strain and did not have a matching major histocompatibility complex. Although some teratomas were detected, their frequency was low.

The lesson of these studies is that many genetic diseases, especially those with a well-known etiology and a single genetic defect, can be addressed with ES cell technology. Because the stem cells have the ability to be passaged indefinitely, they are reasonable targets for genetic manipulation of the sort that produced Poly. Other diseases that fall into this category include familial hypercholesterolemia, severe combined immunodeficiency, cystic fibrosis, and certain cancers.

Other Tissue Failures

There are many other diseases involving tissue failure that may be addressed with stem cell therapy. One of them is sickle-cell anemia, a disease affecting about 80,000 Americans.

This painful and deadly disease is caused by a variant form of hemoglobin. As it oxygenates tissues, the malformed hemoglobin clumps together and forms stiff rods that distort the red blood cells into a distinctive sickle shape. This inflexible blood cell can clog capillaries and starve organs of oxygen, leading to damage and pain. One of the first organs to succumb is often the spleen, leaving the victim susceptible to opportunistic infections.

In January 2006, a team led by Michel Sadelain of the Memorial Sloan-Kettering Cancer Center reported on a successful trial using RNA interference (RNAi) and stem cells to correct the genetic deficits leading to **sickle hemoglobin**. RNAi was introduced with a lentiviral vector to block expression of the aberrant hemoglobin gene in hematopoietic stem cells. The RNAi was combined with the gene for normal gamma globin. This combination both knocks out the bad gene and expresses the proper gene. The cells were then able to differentiate into normal red blood cells, producing a functional hemoglobin molecule while at the same time suppressing the sickle-cell variant. This research demonstrates that stem cells can be reprogrammed to reverse genetic defects without resorting to allogenic grafts and their inherent compatibility problems.

Summary

The existing ES cells cultures have been grown on mouse feeder cells, making them unsuitable for human trials. New techniques have been developed to grow cells in animal-free cultures.

Several diseases are currently targeted for ES therapy. Type 1 diabetes may be treatable with islet cells developed from ES cells. Although much progress has been made, the most robust cells produce low levels of insulin, and the cells producing the most insulin are fragile. Type 2 diabetes is being approached by research involving genetic repair.

Cardiac stem cells have yet to be definitively identified, but research has shown that ES cells can differentiate into at least three of the tissues needed to repair a wounded heart: cardiomyocytes, vascular smooth muscle, and endothelial cells. Liver cells have also been differentiated from ES cells and have been shown to successfully engraft in the host.

Cancer presents new challenges for stem cells. Much of the work has centered on stimulating immune stem cells to better target metastatic cells, typically using bone marrow allografts. Other research has buttressed the notion that many cancers may be propagated by cancer stem cells. These cells may be resistant to current chemotherapy and radiation therapy, possibly explaining many cases of recurrence.

Hemophilia is another disease targeted for stem cell therapy. One of the original motivations for the nuclear transfer experiments that led to Dolly the sheep was to create an animal that could express clotting factor IX in its milk, and this has proven successful. Other experiments have shown that liver cells derived from ES cells can survive engraftment into the liver and produce factor IX.

Many of these experiments demonstrate that ES cells can be reprogrammed to compensate for a genetic defect and then differentiated into multiple cell types. This makes ES cells well suited for treating diseases that are known to be related to genetic defects, such as hemophilia, cystic fibrosis, sickle-cell anemia, and many more.

Review Questions (True or False)

1. Stem cells have been used to treat cancer since the 1950s.
2. RNAi techniques can be used to genetically repair ES cells before amplification and differentiation.
3. Some cancers may arise from rogue stem cells that are resistant to current therapies.
4. So far, it has not been possible to differentiate hepatocytes from ES cells.
5. It is not possible to grow entire organs from ES cells.
6. Pancreatic cells and neural cells are both induced from ES cells by the protein nestin.
7. The 64 stem cell lines approved by President George W. Bush cannot be used in human trials due to mouse cell contamination.

ADDITIONAL READINGS

Assady, S., Maor, G., Amit, M., Itskovitz-Eldor, J., Skorecki, K. L., and Tzukerman, M. (2001). Insulin production by human embryonic stem cells. *Diabetes* 50: 1691–1697.

Chinzei, R., Tanaka, Y., Shimizu-Saito, K., Hara, Y., Kakinuma, S., Watanabe, M., Teramoto, K., Ariil, S., Takase, K., Sato, C., Teradae, N., and Teraokae, H. (2002). Embryoid-body cells derived from a mouse stem cell line show differentiation into functional hepatocytes. *Hepatology* 36: 22–29.

Itskovitz-Eldor, J., Schuldiner, M., Karsenti, D., Eden, A., Yanuka, O., Amit, M., Soreq, H., and Benvenisty, N. (2000). Differentiation of human embryonic stem cells into embryoid bodies comprising the three embryonic germ layers. *Mol Med* 6: 88–95.

Jacobson, L., Kahan, B., Djamali, A., Thomson, J., and Odorico, J. S. (2001). Differentiation of endoderm derivatives, pancreas and intestine, from rhesus embryonic stem cells. *Transplant Proc* 33: 674.

Kay, M. A., Manno, C. S., Ragni, M. V., et al. (2000). Evidence for gene transfer and expression of factor IX in haemophilia B patients treated with an AAV vector. *Nat Genet* 24: 257–261.

Kehat, I., Kenyagin-Karsenti, D., Druckmann, M., Segev, H., Amit, M., Gepstein, A., Livne, E., Binah, O., Itskovitz-Eldor, J., and Gepstein, L. (2001). Human embryonic stem cells can differentiate into myocytes with structural and functional properties of cardiomyocytes. *J Clin Invest* 108: 407–414.

Lanza, R. P., Chung, H. Y., Yoo, J. J., et al. (2002). Generation of histocompatible tissues using nuclear transplantation. *Nat Biotechnol* 20: 689–696.

Lumelsky, N., Blondel, O., Laeng, P., Velasco, I., Ravin, R., and McKay, R. (2001). Differentiation of embryonic stem cells to insulin-secreting structures similar to pancreatic islets. *Science* 292: 1389–1394.

Nathwani, A. C., Davidoff, A. M., Hanawa, H., Hu, Y., Hoffer, F. A., Nikanorov, A., Slaughter, C., Ng, C. Y. C., Zhou, J., Lozier, J. A., Mandrell, T. D., Vanin, E. F., and Nienhuis, A. W. (2002). Sustained high-level expression of human factor IX (hFIX) after liver-targeted delivery of recombinant adeno-associated virus encoding the hFIX gene in rhesus macaques. *Blood* 100: 1662–1669.

Orlic, D., Kajstura, J., Chimenti, S., Jakoniuk, I., Anderson, S. M., Li, B., Pickel, J., McKay, R., Nadal-Ginard, B., Bodine, D. M., Leri, A., and Anversa, P. (2001). Bone marrow cells regenerate infarcted myocardium. *Nature* 410: 701–705.

Osorio, R. W., Ascher, N. L., Jaenisch, R., Freise, C. E., Roberts, J. P., and Stock, P. G. (1993). Major histocompatibility complex class I deficiency prolongs islet allograft survival. *Diabetes* 42: 1520–1527.

Richards, M., Fong, C. Y., Chan, W. K., Wong, P. C., and Bongso, A. (2002). Human feeders support prolonged undifferentiated growth of human inner cell masses and embryonic stem cells. *Nat Biotechnol* 20: 933–936.

Soria, B., et al. (2000). Insulin-secreting cells derived from embryonic stem cells normalize glycemia in streptozotocin-induced diabetic mice. *Diabetes* 49: 157–162.

Theise, N. D., Nimmakayalu, M., Gardner, R., Illei, P. B., Morgan, G., Teperman, L., Henegariu, O., and Krause, D. S. (2000). Liver from bone marrow in humans. *Hepatology* 32: 11–16.

Yamada, T., Yoshikawa, M., Kanda, S., Kato, Y., Nakajima, Y., Ishizaka, S., and Tsunoda, Y. (2002). In vitro differentiation of embryonic stem cells into hepatocyte-like cells identified by cellular uptake of indocyanine green. *Stem Cells* 20: 146–154.

Yin, Y., Lim, Y. K., Salto-Tellez, M., Ng, S. C., Lin, C-S., and Lim, S-K. (2002). AFP+, ESC-derived cells engraft and differentiate into hepatocytes in vivo. *Stem Cells* 20: 338–346.

Human Embryonic Stem Cells and Society

The Religious, Legal, Ethical, and Scientific Debate

Despite many claims to the contrary, life does not begin at conception: It is an unbroken chain that stretches back nearly to the origin of the Earth, 4.6 billion years ago.

Carl Sagan, 1998

OVERVIEW

As discussed in Chapter 12, Dr. James Thomson used surplus blastocysts from a fertility clinic for his stem cell research. The embryos, scheduled for destruction, were donated by the parents for this project. He knew the federal funding for his animal research at the University of Wisconsin could be withdrawn if it overlapped in any way with his privately funded human embryo research. He had taken great pains to separate the laboratories, not sharing so much as an extension cord between the two. It was expensive and redundant but absolutely essential because the federal law was unambiguous in its ban on funding human embryo research, first stated in 1996 and renewed every year thereafter. This ban had originally created an unprecedented void in the United States of basic science research to underlie the clinical treatment of infertility by assisted reproduction, and it now extended to embryonic stem (ES) cell research.

Related to the federal ban on research was the underlying view held by many U.S. citizens that life begins at fertilization, and thus research on fertilized eggs was likened to an early abortion. There is no doubt that elective abortion is a tragedy for all involved. Some antiabortion groups have become extremists and feel justified in killing doctors who perform abortions. They also maintain Internet hit lists that provide names and addresses of doctors and scientists who work in any capacity with human embryos. The irony is that by stopping research on fertilized eggs and early embryonic development, the extremists actually inhibit the scientific understanding of the process and thus the development of new ways to prevent pregnancy and in this way perpetuate the need for abortion itself. There is no substitute for knowledge.

Thus to Jamie Thomson's concerns about federal funding for his nonhuman research were added the additional worries about being on the hit list of radical groups. On the one hand, concern for each and every human life speaks well for the values in our society. On the other hand, when should the concern begin and who should establish the ethical guidelines? These are complicated social issues based on centuries of religious and theoretical dogmas that require much public debate to resolve.

The Religious Debate

The ablation of the inner cell mass (ICM) of the blastocyst, which critically and irremediably damages the human embryo, curtailing its development, is a gravely immoral act and consequently is gravely illicit.

Vatican City, August 25, 2000

Some, but by no means all, religions believe that human life starts at conception. The Roman Catholics are the most powerful and adamant about this concept, but they were not always so. In the 5th century, Saint Augustine proclaimed that abortion before the **quickening**—when the fetus becomes large enough to have its movements felt by the mother, approximately 4 to 5 weeks of development (see Chapter 9)—was neither a sin nor a homicide, because "one cannot say that there is a living soul in a body that lacks sensation . . . the flesh has not yet been formed and thus has no capacity to feel."

In the 12th century, Saint Thomas Aquinas advanced the theory of delayed "hominization," or "ensoulment." Aquinas saw embryogenesis as proceeding through three stages: vegetative, animal, and rational. Only in the ultimate rational stage did the fetus gain a soul and therefore become a human being. As did his

Quickening: the sensation in the uterus caused by movement of the fetus.

predecessors, Aquinas pegged that event at day 40 (counting from the last menstrual period), concluding that abortion before then could not be considered murder.

At the start of the 13th century, Pope Innocent III reiterated that the quickening was still considered to be the start of life and that before day 40, abortion was not homicide. From 1140 to 1917, that was canon law. In the 16th century, Pope Gregory XIV confirmed once again that life starts at the quickening and that abortion before then was not homicide. In 1869, however, Pope Pius IX changed the direction of the church. For the first time in Catholic history, abortion—from the moment of conception—became the equivalent of homicide. The 1917 Code of Canon Law put it all in writing.

What convinced the church that conception was the magic moment when life begins? It was the explanation by scientists that fertilization involved sperm and eggs. Observations since van Leeuwenhoek first spotted sperm with his microscope led to the establishment of embryology in the 1800s. By the middle of the century, it was established that sperm and eggs were both necessary and sufficient for the formation of a zygote, although it was still unclear what parts of these cells contained the hereditary components.

Thus for nearly 2,000 years the church accepted abortion in one form or another, and only for the last 150 years has it been banned, based on the science of that era. As science once again shifts paradigms, the Roman Catholic Church may again reinterpret its canon.

If the history of the Roman Catholic religion is not a picture of unwavering conviction on the issue of abortion, neither is any other religion. Religious thinkers from all major religions have come down on one side or another, at one time or another.

For example, Islamic scholars are not unanimous on abortion—and some imams strictly prohibit it—but the general theory is similar to Aquinas' delayed ensoulment. Islam usually accepts abortion before the 80th day but also believes that the sin increases with the duration of the pregnancy.

Buddhists treat life as a continuum, both in time and between species, so it is possible to consider a bug to be as sacred as an embryo. Abortion is therefore wrong and negatively affects your karma. However, there are rituals to dispel bad karma, so the stain of sin may be temporary. Most modern Buddhists agree that the decision of whether or not to abort belongs to the mother.

Judaism, like Islam, is not unified on the issue of elective abortion, but in general the consensus among Reformed Jews is that abortion is a woman's choice. Although the fetus is still sacrosanct, it is not considered a person until it takes its first breath of life at birth—the Judaic version of delayed ensoulment. The Presbyterian Church supports a woman's right to choose, as do many other American Christian groups.

In summary, although some religions have strong convictions, the opinion of religious scholars is prismatic and fractured. Although most thinking comes down on the side of a mother's right to choose, there is plenty of dissension. Abortion is an issue that affects men and women differently, and abortion rights often track advances in women's rights. As liberal governments have flourished, human rights have expanded, and today about two-thirds of the world's population lives in countries where women have at least nominal rights and abortion is legal.

Religious opinions of ES cell research closely track positions on abortion. The religions that are liberal on abortion are generally positive about stem cell research.

For example, The Church of Scotland "affirms the special status [of the embryo] as created by God, but it recognizes the potential benefits of embryo research under limited circumstances." It approves studies with nuclear transfer as well but hopes that its use will be limited and aimed at ultimately creating therapies that remodel somatic cells without using an egg.

Bert Clifford, the chair of the Ethics Committee at Sydney Adventist Hospital, says a ban on stem cell research has no moral justification. "We should be open to dialogue, open to discussion, open to possibilities, open to finding the best solution in changing times in advancing technology."

The Episcopalian Church is divided, but some bishops claim that embryonic research should be permissible on embryos up to 14 days old. The Presbyterian stance is also favorable and includes the donation of embryos from fertility clinics. The Church has issued a statement saying that "Prohibition of the derivation of stem cells from embryos would elevate the showing of respect to human embryos above that of helping persons whose pain and suffering might be alleviated. Embryos resulting from infertility treatment to be used for such research must be limited to those embryos that do not have a chance of growing into personhood because the woman has decided to discontinue further treatments and they are not available for donation to another woman for personal or medical reasons." The Union of Orthodox Jewish Congregations expressed similar sentiments in 2001, saying, "an isolated fertilized egg does not enjoy the full status of personhood."

At the other end of the spectrum, the Lutheran and United Methodist Churches agree with the Vatican and are opposed to human ES cell research. Similarly, the Southern Baptist Convention asserts, "Human embryos are the tiniest of human beings."

Another impassioned voice against human ES cell research is the American Life League, which grounds its theocracy in Roman Catholic teachings. The American Life League bills itself as the largest grass-roots pro-life organization in the United States. They want to criminalize ES cell research, and they also want to ban fertility clinics and contraceptives.

The erratic history of religious pondering on the topic of when life begins indicates that the topic should not be considered closed. The forgiveness extended to both Galileo and Darwin by the Roman Catholic Church demonstrates that the Roman Curia can change its mind and accept scientific evidence. Although the Pope may be considered to be infallible, no such claim is made for Catholic law, which is allowed, and expected, to change.

Science has confirmed Saint Augustine's observations that the "flesh" of an early embryo is indeed without form and bereft of sensation. Perhaps the most curious aspect is that despite the high esteem held by the church for both Aquinas and Augustine and the fact that modern science seems to be vindicating their observations, the opposite conclusion is being currently propounded by the Vatican.

The religious argument against ES cell research would be much stronger if all religions agreed. Instead, not only do many religions find the research to be permissible, but some even believe it would be immoral not to use stem cell therapies. The problem is difficult, and emotions run high. Religions are attempting to reconcile the overlapping rights of three parties: the mother, an embryo, and the patient. The answer is not an easy one.

The Legal Debate

The Department of Health and Human Services (DHHS) concluded that the Congressional prohibition [on the funding of embryo research] does not prohibit the funding of research utilizing human pluripotent stem cells because such cells are not embryos.

National Institutes of Health Guidelines for Research
Involving Human Pluripotent Stem Cells, 2001

In many countries, the legal disposition of abortion is dictated by religious leaders. Most Western countries, however, have attempted to separate religion and law. Modern societies, where travel and foreign trade flourish, find it difficult to impose any single religion on diverse citizenry. Secular civil law, based on democratically elected governments, has proven to be a successful formula for economic and scientific progress. Unfortunately, leaving religion out of the legal system has not really simplified matters. The legal standing of an embryo has been in constant flux throughout American history. It has variously been defined as a human, as property, and, more recently, as a new type of entity that is not owned but whose disposition belongs to the gamete donors.

In the 1930s, researchers routinely used fetal tissue from abortions and miscarriages for experimentation and, most importantly, as an invaluable source of cultured cells for studying bacteria and viruses. In 1949, it was a human fetal kidney cell culture that finally allowed Harvard researcher John Enders to grow the polio virus in vitro. This work ultimately led to the development of the polio vaccine by Jonas Salk and Albert Sabin. For this life-saving breakthrough, Enders was awarded the 1954 Nobel Prize for Medicine.

In 1973, ignoring the strongly stated objection of the Roman Catholic Church, the U.S. Supreme Court ruled in Roe v. Wade that a woman has a legal right to terminate her pregnancy up until the point where the embryo is viable and able to survive on its own outside the womb. The court concluded that a fetus is not a person within the meaning of the Fourteenth Amendment. Thus, the idea that a 5-day-old blastocyst is a human being is not supported by U.S. law.

Nonetheless, the emotions inflamed by the Roe v. Wade decision led to a moratorium on federally funded embryo research. The following year, Congress established the National Commission for the Protection of Human Subjects of Biomedical and Behavioral Research, which sought to regulate embryo research.

In 1987, the National Institutes of Health (NIH) placed a moratorium on federal funding of all human embryo research until an advisory panel could issue guidance on the types of research that should be permitted. In 1988, the panel voted 19–2 in favor of funding for human embryo research with the proviso the projects be reviewed by an ethics advisory board. But, this recommendation not withstanding, Louis Sullivan, the Secretary of the DHHS at the time, sided with the two dissenters and continued the moratorium by not convening an ethics advisory board. Subsequently, the U.S. Congress twice voted to override the DHHS moratorium, but then President George Bush the elder vetoed the votes. The ban on federal funding stayed in effect.

In 1993, during the presidency of Bill Clinton, Congress passed the NIH Revitalization Act. This Act provided a regulatory framework for research on fetal and embryonic tissues. The tissue would have to be accompanied by written permission from the mother or donor affirming that the donation was for therapy research, that no restrictions would be placed on the recipient, and that the recipient would remain anonymous. The Act also required that the mother make the decision to abort before and separately from the decision to donate tissue. This was to ensure that the option of donating embryos for research did not encourage women to choose abortion. There is, however, no known correlation between embryo research and abortion, anymore than there is between using the organs of murder victims and the rate of murder.

In 1994, another panel convened to advise Congress on the issue of human embryo research again voted in favor of research. The Revitalization Act had eliminated the proviso that research on fertilized eggs and early embryos needed to be reviewed not only by NIH peer review committees but also by the ethics advisory board, which had never been convened. This opened the possibility that research on fertilized human eggs could be funded by NIH, but two congressmen were opposed to such research going forward without congressional debate. As a consequence, they attached a rider to the NIH budget in 1996, known as the Dickey-Wicker Amendment, that specifically denied NIH funds to be used for research on human embryo or embryos. (For purposes of this section, the term *human embryo or embryos* includes any organism, not protected as a human subject under 45 CFR 46 as of the date of the enactment of this Act, that is derived by fertilization, parthenogenesis, cloning, or any other means from one or more human gametes or human diploid cells.)

The Dickey-Wicker amendment to budget appropriations for the NIH has been renewed every year by Congress. These regulations govern federally funded research, but they do not carry criminal sanctions. There are, however, some explicit laws governing what can be done with embryos and/or fetal tissue. It is against the law to profit from interstate commerce in embryos, it is illegal to solicit or accept donated tissue when the recipient is a friend or a relative, and it is illegal for the recipient to pay for the procedure. In all cases, the ultimate arbiters of the disposition of the fetus are the parents.

At first, President Bill Clinton promised to remove the Dickey-Wicker amendment, but after a well-organized write-in campaign by Christian groups, including the Roman Catholic College of Cardinals, then President Clinton decided not to set aside the ban on federal funding.

It is important to note that although federal funding for embryo research is unavailable from the government, the research itself is not illegal. This distinction has been lost on many in the anti–stem cell movement, including Senator Sam Brownback, who has written that "embryonic stem-cell research is illegal, immoral and unnecessary." In fact, it was and still is completely legal to do embryo research with private financing. It was private funding from Geron that allowed James Thomson to isolate the first human ES cells.

President George W. Bush's declaration, on August 9, 2001, that he would allow stem cell research on 64 existing embryonic cell lines, opened the lid slightly on federal funding. He stated, "I have concluded that we should allow federal funds to be used for research on these existing stem cell lines where the life and death decision has already been made." Unfortunately, these cell lines—for reasons relating to the infancy of the research—are of uncertain utility. Some are not expanding properly, and others seem to have lost their capacity to differentiate. All the lines have apparently been grown on mouse fibroblasts, making them unsuitable for human therapy. Comparing data on these cell lines is complicated by the fact that these cells were created in independent research laboratories from around the world, each following separate protocols.

However, that decision is rightly the parent's to make, not the government's. There are hundreds of thousands of frozen embryos left over from fertility procedures around the world that could be used to create new ES cell lines using better standardized procedures. These cell lines, like blood and bone marrow, are most useful when there is a tissue type match. That is a powerful argument in favor of more ES cell lines, not just the limited set proposed by the President.

The number that have been made available, along with the red tape that it takes to procure them, compounded further by conflicting patent rights and ownership claims, has resulted in very few federally funded ES cell studies. As of October 2002, only four stem cell material transfer agreements had been signed. The NIH lists 34 stem cell lines as being eligible for federal funding in early 2006, but most of the lines are effectively unavailable to researchers.

As disappointing as these results are, they nevertheless point to a loosening of the government's ban on embryo research, hastening the promise of eventual therapy. However, before many of the NIH grants go out, the Congress may ban certain aspects of this research entirely. On August 1, 2001 the House of Representatives voted overwhelmingly (265–162) to make nuclear transfer using human cells a crime, punishable by a million dollar fine and 10 years in jail. This would have become the law of the land if the Senate version of the bill (SB-790) had also passed. Neither bill is currently active, although new legislation is regularly introduced.

Denying federal funding has set back embryo research for over a decade. The sponsors of the ban hope to put such a chill into the field that they can stop it entirely, forcing such research to move overseas.

That is happening already. Several researchers have left the United States to start new research laboratories in England, Singapore, Israel, Saudi Arabia, and Japan, where the laws are more liberal and the governments are promoting the science. Each of these countries expects to benefit greatly from American intransigence.

Moving in the opposite direction, in September 2002 then California governor Gray Davis signed a law explicitly promoting stem cell research, including nuclear transplant stem cells. In 2004, under the governorship of Arnold Schwarzenegger, California made national news by overwhelmingly approving Proposition 71, a 10-year three-billion-dollar, stem cell initiative. However, progress in California has been stymied by various lawsuits claiming that the research is unethical and institutionalizes several conflicts of interest. In April, 2006, California superior court judge Bonnie Lewman Sabraw ruled against the lawsuits and in favor of the stem cell initiative, thus clearing the way for California stem cell research to move forward.Some scientists worry that the California initiative takes the pressure off the Federal government to propose its own programs, but it at least signals a change in people's opinions on the utility of stem cell research to end the suffering of millions of Americans.

The Ethical Debate

> [Researchers] should be asking who this research should be done on, will it work, what will the cost be and who should have access to it. But because of abortion politics, we spend no time at all talking about this.
>
> Arthur Caplan, University of Pennsylvania Center for Bioethics

The ethical debate is in the eye of the storm about human ES cells, and those attempting to make calm judgments do so as the windy arguments between religion, the law, and science churn around them. Bioethics committees are an important part of any research project. Most scientists welcome their recommendations, because it lets them proceed with a minimum of moral ambiguity.

The first bioethicist of record was Hippocrates, the author of the injunction to "first, do no harm." For over 2,000 years, his paternalistic vision prevailed and helped medicine to become a professional almost priestly trade. Doctors, it was believed, could be counted on to produce wise and moral judgments about a patient's health—judgments that were considered too difficult for the patient to possibly comprehend. But this comfortable relationship between the childlike patient and the authoritative doctor-minister was not to last.

The end of World War II exposed the depredations of Nazi doctors who performed experiments on unwilling human subjects. In response to these horrors, the Nuremberg Code was created to establish a list of human rights that must be observed by all moral and ethical research. These principles are summarized as follows:

1. The voluntary consent of the human subject is essential.
2. The experiment should yield fruitful results for the good of society, unprocurable by other means.
3. The experiment should be based on the results of animal experimentation, and the anticipated results should justify the experiment.
4. The experiment should avoid all unnecessary physical and mental suffering and injury.
5. No experiment should be conducted where there is a reason to believe that death or disabling injury will occur.
6. The risk to be taken should not exceed the problem addressed by the experiment.
7. Proper preparations should be made to protect the experimental subject.
8. The experiment should be conducted only by scientifically qualified persons.
9. The human subject should be at liberty to end the experiment.
10. The scientist must be prepared to terminate the experiment if he or she has cause to believe that the experiment is likely to result in injury to the experimental subject.

For persons who believe that blastocysts are human beings, stem cell research is clearly unethical on these grounds. For persons who do not believe a blastocyst is a human, the second clause, which requires the consideration of other means of achieving the same goals, such as adult stem cells, remains problematic. Otherwise, stem cell research satisfies these guidelines.

The Nuremberg Code was, unfortunately, not enough to stop the infamous Tuskegee Syphilis Study in Alabama. This research project, approved and reapproved over a 40-year period by the U.S. Health Services, denied poor black men treatment for syphilis, even though the cure (penicillin) was discovered during the trial. The Tuskegee project not only brought well-deserved opprobrium to the Health Services, and much stricter rules about informed consent, it proved that federal institutions should not be considered the ultimate authority on ethical guidelines for research.

Institutional review board: the institutional committee that enforces guidelines for research involving human subjects.

Bioethics committees now abound. Most universities have an **institutional review board** or a research ethics committee to clarify the legal and ethical issues involved with any kind of human research. Private companies have set up bioethics committees composed of legal, religious, and ethical scholars. Capping it all off is the National Bioethics Advisory Commission established by President Clinton in 1994 and continued by President George W. Bush.

Not surprisingly, ethicists are about as unified as religious scholars on the merits and dangers of ES cell research. However, their view of a human almost always includes sentience or at least a history of sentience. On that basis, most ethicists do not consider a blastocyst to be a person. An argument can also be made that the blastocyst is not destroyed, because the cells that were destined to become the embryo are exactly the stem cells that are cultured. Because these cells are still alive and because they are still in an undifferentiated state, in what sense has the blastocyst been killed? One cannot even say it is "dismembered," because there are no members in a blastocyst. However, if full personhood is accorded to a blastocyst, it would still be objectionable to disassociate the cells, even if death were not the result. Nevertheless, the notion that a blastocyst could be disassembled and then reassembled—and that its cells are immortal—puts it into a special ethical category distinct from ordinary mortal human beings.

An ethos that protects the human right to privacy could view ES cell therapy as a simple private arrangement between, say, spouses who wish to fertilize an egg to create stem cells for their own therapy. If spouses are granted the privacy to make children and they are not required to report miscarriages to the authorities, then would it be ethical to create a blastocyst that might have therapeutic qualities? If this level of privacy is not a right and if blastocysts are human beings, then all pregnancies would need to be reported to some legal agency that would track them to ensure that each one comes to term and is not diverted for therapeutic purposes.

A simpler ethical decision is whether or not to use surplus embryos beyond the ones already specified by President George W. Bush. There are hundreds of thousands of frozen blastocysts in fertility clinics around the world, and the vast majority of them are destined to be destroyed. The simple ethical conclusion is to use them. In fact, most scholars believe that it is unethical to let sick people die by denying them a potential treatment. That is, after all, what happened in the Tuskegee study.

The argument against using the surplus blastocysts is that it might encourage people to create embryos for therapy. However, the law prevents people from profiting from the sale of embryos, so the only reason to create a therapeutic embryo would be compassion. However, because some Christian groups do not call it compassion but abortion, it is the job of the ethicist to try to reconcile these two diametrically opposed views.

A starting point for many ethicists is the utilitarianism of Jeremy Bentham and John Stuart Mill, who argued that the best civilizations perform "those acts which result in the greatest good for the greatest number of people." The point is to maximize happiness and minimize pain for all involved. If those involved include an insensate blastocyst, not yet capable of feeling either happiness or pain, and a sick patient who could benefit from that tissue, then the conclusion is easy: Sacrifice the blastocyst for the good of the sick patient. To do otherwise would be immoral.

Utilitarianism is believed by some scholars to be a thinly disguised justification for capitalistic excess, which they fear will start mankind down a slippery slope to merchandizing human beings. Utilitarians like John Stuart Mill were dead set against such slavery, which they believed could hardly maximize happiness—but scoundrels have been known to take refuge behind utilitarian arguments. A more serious problem most religions have with utilitarianism is that it only counts sentient creatures—if something can feel neither happiness nor pain, it is simply not a part of the equation. It is a "nonmoral" agent, bereft of human rights.

Utilitarianism is a laudable attempt to quantify and calculate units of pleasure and pain, but it is far from an exact science. Even in ordinary cases, it poses some difficult questions. In extraordinary cases, like the quagmire of ES cell research, it is merely a starting point for the discussion.

Stem cell ethics committees generally consider several overlapping concerns when they review a research project:

- The laws governing human embryo research;
- The scientific goals and techniques of the research;
- Potential applications of the research;
- The procedures by which human tissues are obtained;
- Federal funding restrictions regarding human embryo research;
- Alternative methods that might achieve the same goals;
- The value of scientific freedom;
- Public opinion;
- The recommendations of other committees.

Ethics committees often engage religious scholars, with the understanding that multiple views should be expressed. However, certain of these spiritually oriented thinkers, although well versed in religious philosophy, have little special education in modern reproductive and medical therapies. Some ethicists believe that although including the views of this segment of society is important, it is not wise to give them a formal position because it institutionalizes the very conflict of values that church–state separation was designed to solve.

Others, such as Robert Veatch, believe that expert groups cannot be representative of a modern pluralistic society. Although this is usually true, it ignores an important problem: Without a full presentation of the facts, how are people expected to make informed decisions? And, after the facts have been delivered, should not one hope that at least some expertise has been gained?

Ethics committees have been blasted for being biased, for having too many religious people or too few, for being beholden to a corporate philosophy, and for many other perceived or real sins. However, the mere existence of these committees is a positive sign that researchers, governments, and corporations are all looking to do the right thing.

The Scientific Debate

Most of us in the field think the push to define adult or embryonic cells as better than the other is a lot of folly.

George Daley, Harvard Medical School

If scientists could create ES cells without using an egg, they would. However, at this time other ways of creating ES cells are not known. As described in the preceding chapters, before a woman's eggs are fertilized, they each have their own identity, defined by the unique genetic mixtures of maternal and paternal genes that occur during meiosis (see Chapter 3). The woman herself was once a tiny egg in her mother's ovary, a unique mixture of her maternal grandparent's genes—and so on before that, in an unbroken chain, back through each maternal egg to the first cell on the face of the earth, several billion years ago. Like the Buddhists, biologists share a fondness for continuity in life.

The moment when a sperm and an egg meet provides a romantic starting point for life, but in reality that moment is pretty fuzzy. As described in Chapter 5, there are dozens of steps involved in fertilization, but at that first meeting there is no blending of male and female genes. Therefore, if combining DNA is a signal event, it cannot be said to happen at the moment of fertilization. Therefore, from the biologic point of view, conception fails as a signpost for novel life.

It is also clear that not every zygote is destined by nature to complete the journey. Along the way, problems may occur that cause up to two-thirds of all zygotes to be lost. Some religions prohibit abortion because they believe it is God's plan that each conception should lead to an infant. However, the natural path from conception to birth is not so assured. Until about day 14, the blastocyst has the potential to split and form twins. The idea of personhood before this time is counterintuitive because we are not accustomed, mentally or legally, to seeing people split in two. If the original blastocyst had full human rights, could it sue its cloned twin for stealing half of its "body?" Until the potential of becoming a unique human is actually manifest in the embryo, we cannot confer humanhood without provoking paradoxes.

The 14-day-old embryo can split because it has no organized tissue to dismantle. There is no hint of a nervous system yet, either centralized or peripheral. For these reasons, several governments, including England's, allow research on human embryos up until the second week, when the "primitive streak" is formed (see Chapter 9).

Scholars since Saint Augustine have assumed that sentience is delayed in the developing embryo. Biology has confirmed these assumptions: The neural tube forms between weeks 3 and 4 of embryonic development (see Chapter 9), and by week 8 the brain starts to take form. By the third trimester, brain waves can be recorded. Might sentience be a useful marker for the start of life? This would bookend well with the end of life as it is viewed by most religions. Consider the following.

When organ and tissue transplants were first proposed in the 1950s, all major religions came out forcefully against them. Transplants from dead people into the living were something straight out of Frankenstein. But within a few years—as the prognosis for transplant recipients improved—attitudes softened. Today, all major religions, with the notable exception of Shinto, have accepted organ transplants as a moral therapy. One of the things they insist on is that the donor be dead before essential organs are harvested. Death is defined as flat-line brain waves. In other words, the major religions consider that humanhood ceases upon brain death. This seems like a reasonable dividing line; we instinctively know that without a brain, the person is gone.

If mental function is essential to humanhood and its extinction sufficient to prove "nonhumanhood," then it might be reasonable to assume that humanhood also starts with mental function. This provides a bright line—that point in development where brains are developed and senses are awakened—that could determine when life begins. This has the evidence of biology to back it up and is buttressed by the appeal of common sense. According to polls, this is what most Americans believe.

A vague consensus, however, is no reason for scientists to be cavalier. They know they are studying life, and that life is sacred. Undeniably, scientists are generally less squeamish when it comes to exploring life than the general population. They know how important it is to learn and to observe from nature. They know that diseases cannot be cured without observation. It is clear, therefore, that dead embryos need to be dissected to discover how to cure birth defects. The lay person may not want to know all the details, but they are generally grateful for the curative therapies that result.

The public has been led to believe that there is a scientific conflict between research on adult stem and ES cells. This is important because if adult stem cells, which do not carry so much ethical baggage, could work, then there would be no need to study ES cells. However, most scientists do not frame the debate in those terms, and most researchers insist that both lines of research are needed.

Neil Theise, an adult stem cell researcher at the New York University Medical School, says, "Losing the opportunity to study embryonic stem cells could mean that we could lose out on some of the enormous therapeutic potential offered by these cells." Catherine Verfaillie, who has made several discoveries about adult stem cells, says, "My message has always been, even though we're excited about the adult cells, it's too early to say that they will replace embryonic stem cells." She points out that much of her research has depended on discoveries made in the ES cell field, and she recommends continuing research on both.

In September 2002, Irving Weisman and Amy Wagers used fluorescent tags to show that adult stem cells can be converted to blood stem cells, but they found virtually no other stem cell products. Wagers concluded that adult stem cell plasticity is not "as robust as it is claimed to be." Other work by Weisman and Wagers has shown that hematopoietic stem cells circulate through the various tissues of the body, which may lead to their misidentification as organ-specific stem cells. These results still need to be reconciled with other experiments that demonstrate greater plasticity of adult stem cells.

The rules of science demand objectivity, which ultimately weeds out religious or political agendas. Scientists will decide which cells to use for which therapies based on experimental findings, but only when those experiments are allowed.

The Future of the Debate

Science faces—in ethics and religion—its most interesting and possibly humbling challenge . . . while religion must somehow incorporate the discoveries of science to retain credibility.

Edmund O. Wilson, Consilience

As James Thomson can attest, each advance in cloning and stem cell technology has been greeted by howls of protest. The usual concern is that the scientists are trying to play God but are instead creating monsters using dark and mysterious technologies. Unfortunately, the language of the stem cell investigator is still baffling to most citizens. Instead of scientists valiantly struggling to cure disease, many see Dr. Frankenstein tending his embryo farm. Scientists, having made their research inaccessible to the lay person, must shoulder some of the blame for the confusion.

To date, opposition to human ES cell research in the United States has been effective. When the American Heart Association offered to support and even fund certain stem cell projects, a mail-in campaign organized by the Roman Catholic Archbishop of St. Louis compelled the American Heart Association to rescind their offer. The same sequence of initial support and then abrupt disavowal occurred with the American Cancer Society and the American Red Cross after similar write-in campaigns.

The result is that the research has been driven into the private sector and overseas. Companies in the private sector depend on patent royalties and trade secrets, and they are not always eager to publicize the results of their research. The ironic upshot is that the research may be driven back into the dark, where there is no federal oversight, no standardization, and possibly no data sharing. This is probably not what most of the letter writers were hoping for. Religion, for all its passion, cannot decide the issue for us. There are simply too many diametrically opposed theologies—many claiming to be the actual word of God—to reach a consensus.

Fortunately, in democratic societies these decisions are up to the citizens and their Constitutions. Unfortunately, many of the people feel ill equipped to render an informed decision. Science is hard to understand, religion offers conflicting advice, the law is a muddle, and ethics are too abstract. In the face of all that, about two-thirds of Americans polled still favor ES cell research.

Into the breach created by lack of federal oversight, the National Academies of Science (NAS) has weighed in with a set of guidelines that bring some welcome structure to human embryonic stem cell research. Since 1863, the NAS has operated under a congressionally granted mandate to advise the federal government on science and technology. In 2005 they created a list of

23 recommendations that address the most sensitive aspects of this research. Among them are the following:

- For each institution involved with human ES cells, an Embryonic Stem Cell Research Oversight (ESCRO) committee should be established. This committee is not expected to substitute for an Institutional Review Board (IRB), but to add an extra level of review. ESCRO committees should include relevant medical experts, ethicists, legal scholars and representatives of the public.
- Each institute should document the provenance of every cell line used and maintain a registry of investigators and their research. This recommendation applies to banked cells as well.
- Research should not involve culturing human embryos for more than 14 days and should not introduce human ES cells into nonhuman primate blastocysts. In addition, animals with transferred human ES cells should not be allowed to breed.
- The scientific community should ensure maximal genetic diversity to provide potential therapy to all groups equally.
- Informed consent must be granted by all donors of gametes, blastocysts or somatic cells for such research. Full disclosure of the goals of the research should be provided, including an understanding that the embryos will be destroyed in the process of collecting the ES cells.
- The confidentiality of the donor must be protected and researchers are warned not to influence the decisions of the donors. Decisions to donate must have no effect on the quality of subsequent care.
- The creation of excess IVF blastocysts for donation should not be encouraged or paid for.
- Women providing eggs for research should only be reimbursed for direct expenses as determined by an IRB.
- Guidelines pertaining to recombinant DNA should also be followed.
- These recommendations should be periodically reviewed.

These are sensible guidelines, and most scientists have welcomed them. They provide a good basis for engendering promising research without impinging on the autonomy or privacy of donors. The requirements for full disclosure are in keeping with the sentiments of the scientific community at large, which wishes to proceed with research in the full light of day, with maximum transparency.

The NAS report also makes the valuable point that good protocols and record-keeping will encourage efficient sharing of ES cell lines. Recognizing that not all researchers have the resources to create their own cell lines, they note that these recommendations will ultimately lead to broader availability of ES cells to more scientists. The tradition of sharing information and materials with colleagues is a hallmark of good science, and demonstrates to the lay public that expanding knowledge and improving the human condition are the paramount goals. The lack of Federal funding greatly hampers this communal effort.

In any case, the public needs much more education on the biology of stem cells, and many ethical issues remain to be solved.

The answer, ultimately, may come down to "common sense," which dictates to most citizens that a dot of cells is not a human being and that it is wrong to let people suffer when there may be a cure at hand. This prediction is based on the long history of scientific progress and its attendant backlashes. Once a science yields fruit and starts to improve the lot of the average person, the backlash melts away.

That said, a smarter more compassionate plan would be to create a War on Cancer approach to speed the development of ES cell research. Its funding should be in line with its incredible promise. At the same time, strict protocols should be established to ensure that infants are not being conceived merely as fodder for research and that experiments are vetted by ethics committees. The subject is too important to be left unobserved and unregulated.

Today's federally funded scientists depend on peer-reviewed grants and publications. Their work is scrutinized by their peers, the public, the press, the government, and the church. There is no place to hide. By defining areas of research that have been cleared of moral minefields, guidelines can actually be liberating to researchers. The uncertainty of the current regulatory environment is discouraging investment, both economic and educational, in this profoundly important science.

The people with the most to gain are those patients who are sick or dying from one of the diseases targeted by stem cell therapy. For them the long-windedness of the debate is often unbearable. The public would be best served by the approval of federal funding, which will help scientists learn more about the miracles of embryogenesis and fertility, while at the same time providing possible cures for heart disease, Parkinson disease, arthritis, Alzheimer disease, diabetes, and more. In the bargain, federal funding will provide stringent guidelines for the research, and ensure that the government-funded results get the widest possible exposure. For the sake of the millions who could benefit from stem cell therapy, this issue demands resolution—and quickly.

Summary

Although President George W. Bush allowed research on 64 existing stem cell lines in 2001, many of these cultures proved to be unviable and all were grown on mouse feeder cells and hence unsuitable for human therapy. Meanwhile, sporadic congressional threats to criminalize the research have cast a pall over the field, convincing many leading scientists to leave the country to continue their work.

Ethical scientific conduct has continued to mature since the time of Hippocrates, and the Nuremberg Code made explicit many of the most important aspects of research and patient rights. However, these ethical guidelines do not define the beginning of life and hence fail to clarify the moral standing of ES cell research.

Institutional review boards have been established at most universities in an attempt to codify and adjudicate ethical research, and most scientists welcome the guidelines. Although there has yet to be a consensus on what is ethical, the establishment of these review boards is generally seen as a good start.

Scientifically, there are many problems with determining the beginning of human life. From conception to at least day 14, there are issues (such as twinning) that make it difficult to assign novelty or identity to the blastocyst. An argument can be made on both religious and medical grounds that sentience is the actual beginning of life, but this is not a universal sentiment. Although adult stem cells evoke less contention than ES cells, most scientists believe both lines of research are needed.

Ironically, the lack of federal funding only means that there is less government oversight and the risk of less transparency in the resulting research. To mitigate these shortcomings, the National Academy of Sciences committee has crafted an extensive set of guidelines designed to promote the highest research standards and encourage scientific cooperation.

Review Questions (True or False)

1. Most religions agree that life starts at conception.
2. Most religions welcomed the first organ transplants.
3. Adult stem cells are just as pluripotent as ES cells and should be the primary focus of stem cell research.
4. ES cell research is illegal in the United States.
5. Federal funding for embryo research has been banned since 1987.
6. Utilitarianism believes that good governments will promote the greatest good for the greatest number of people.
7. The National Academy of Sciences recommends that stem cell researchers fully document the provenance of each and every cell line they use.
8. The stem cell lines approved by President George W. Bush were grown on mouse feeder cells and cannot be used for human therapy.

ADDITIONAL READINGS

Green, R. M. (2001). *The Human Embryo Research Debates*. New York: Oxford University Press.

Green, R. M., DeVries, K. P., Bernstein, J., Goodman, K. W., Kaufmann, R., Kiessling, A. A., Levin, S. R., Moss, S. L., and Tauer, C. A. (2002). Overseeing research on therapeutic cloning: a private ethics board responds to its critics. *Hastings Center Report* 32: 27–33.

Holland, S., Lebacqz, K., and Zoloth, L. (2002). *The Human Embryonic Stem Cell Debate*. Cambridge, MA: The MIT Press.

Kiessling, A. A. (2004). What is an embryo? *Connecticut Law Review*, 36: 1051–1092.

Kiessling, A. A. (2004). What is an embryo? A rejoinder. *Connecticut Law Review*, 37: 1–23.

Committee on Guidelines for Human Embryonic Stem Cell Research, National Research Council. (2005). *Guidelines for Human Embryonic Stem Cell Research*. Washington, DC: The National Academies, 178 pp.

Glossary

abembryonic: The trophoblast cells directly opposite the inner cell mass of a blastocyst.

acetylcholine: A neurotransmitter in vertebrate nervous systems found both in the brain and in peripheral nerves.

acute disseminated encephalomyelitis: An acute onset inflammation of the spinal cord and brain thought to be brought about by infection or vaccination.

adrenoleukodystrophy: An X-chromosome-linked, severe adrenal gland insufficiency in boys characterized by progressive blindness and mental deterioration.

adrenomyeloneuropathy: A chronic adrenal gland insufficiency in adult males.

allele: The gene on each of two chromosomes in a pair, one inherited from the father and one from the mother; different mutational states in the same gene within the organism and/or within the gene pool for the species.

allogeneic: Pertaining to the same gene pool, but not the same organism.

Alzheimer's disease: A disease involving death or injury of brain cells characterized by memory loss and dementia.

amnion: The membranes arising from polar trophoblast cells that completely encase the developing fetus.

amyotrophic lateral sclerosis: A disease of the motor tracts of the spinal cord, also called Lou Gehrig's disease.

androgenote: A zygote, and the resulting cells that arise following cleavage, that has been manipulated to contain two male pronuclei instead of one male and one female pronucleus.

anemia: A medical condition characterized by abnormally low levels of hemoglobin in circulating blood.

anencephaly: Developmental failure of the brain and spinal cord.

aneuploid: Possessing a chromosome number that is not characteristic of the species.

aplastic anemia: A medical condition characterized by low levels of hemoglobin in circulating blood resulting from decreased production of red blood cells.

assisted reproduction: Procedures performed by health care providers to increase the possibility of pregnancy.

assisted reproductive technology: The medical and laboratory procedures developed to increase the possibility of pregnancy through manipulations of eggs and sperm in laboratories.

astral spindle: Protein microtubules attached at one end to chromosomes and at the other end to a tight organizing region for mitosis.

astrocytes: Star-shaped glial cells that are numerous in the brain and the spinal cord that regulate the extracellular environment; they also respond to injury and may be a type of adult neural stem cell.

asymmetric cell division: The nonrandom distribution of chromosomes during cell division that segregates chromosomes with the most ancestral DNA strands preferentially into one nucleus.

autocrine: The process whereby a cell produces a substance that brings about stimulation of one of its own receptors; the usual circumstance involves secretion by the cell of a hormone or growth factor that then binds to a receptor on the cell's own plasma membrane.

autosomal dominant polycystic kidney disease: Caused by a mutation in either of two genes, *PKD1* or *PKD2*, which encode polycystin 1 and 2.

bilaminar germ disc: An advanced inner cell mass stage of embryonic development that appears around the time of implantation; it is characterized by a layer of endodermal cells adjacent to the blastocoel and a layer of ectodermal cells adjacent to the developing amniotic cavity.

blastocoel: The fluid-filled space inside the blastocyst.

blastocyst: A ball of trophoblast cells sealed together with tight junctions that contains the inner cell mass of undifferentiated embryonic stem cells that gives rise to the embryo.

blastomere: A large, undifferentiated cell resulting from cleavage of an activated egg.

bone marrow: The population of blood and immune stem cells that resides in the spongy areas inside bones, especially long bones.

bone marrow transplantation: The medical procedure involving transplantation of cells from the bone marrow to a recipient with the goal of regenerating healthy blood and immune systems within the recipient. Bone marrow may be transplanted as an allograft from a donor or autologously from the patient.

cAMP: Cyclic adenosine monophosphate, a common intracellular second messenger produced by enzymes that reside in or near the plasma membrane of the cell in response to binding of a growth factor or hormone to the cell membrane; a common target for cAMP is activation of protein kinase A.

cardiomyocyte: A general term for cells contributing to heart muscle.

cardiomyocytes: Stem cells committed to becoming heart muscle cells.

caudal: A position near the tail or hind part of the body.

cell: The smallest living unit of mammals and most plants and animals. Cells are defined by an outside (plasma) membrane and contain a full complement of genes characteristic of the species plus a mixture of proteins and other molecules.

cell differentiation: The transition of a cell to the state of gene expression that allows it to carry out the functions specifically required of its tissue or organ of residence; differentiation is frequently accompanied by changes in the appearance of the cell.

cell division: A process that includes replication of all the DNA in each chromosome to create a second set of chromosomes that are then divided equally into the two halves of the cell, which pinches the plasma membrane in the middle to create two new, identical daughter cells.

cell fusion: The process of two cells coming together through fusion of their plasma membranes.

central pontine myelinolysis: Loss of myelin in the brainstem as a result of malnutrition.

cephalic: Pertaining to the head or brain.

cerebral palsy: A disorder of muscular coordination and speech caused by brain damage before, during, or shortly after birth.

chimera: An organism containing two or more genetically distinct cells. An example is a mixture of blastomeres from two mouse embryos that form one blastocyst, which gives rise to a mouse with tissues that contain daughter cells from both mouse embryos.

chorionic villous membrane: A trophoblast cell derivative that forms the interface between the developing embryo and the lining of the uterus.

chromosome: A collection of genes contained within one long, double strand of DNA, which condenses into a distinct entity during cell division.

chromosome pairing: The process of alignment of pairs of matching chromosomes in such a way that one chromosome from each pair will divide with each daughter cell.

clonal expansion: The repeated division of a single cell into a large population of identical daughter cells.

clone: Greek for twig, which can be stripped from a tree and planted to grow an identical tree; used as a scientific term to mean genetically identical.

cloning: Cloning is the act of creating a clone, either as single cells or as individuals. For example, twins are clones of each other, and a cow arising from nuclear transplantation into an unfertilized egg is a clone of the donor of the nucleus.

compaction: The process of blastomeres forming bonds between their plasma membranes that results in adherence of the cells together.

constitutive expression: The continual expression of a gene into its protein product without the need for specific, outside stimulus.

corona radiata: The term used to describe the microscopic appearance of a layer of cells next to the matured egg that are spread out from touching each other by the synthesis of the glycoprotein hyaluronic acid.

cumulus cells: The cells surrounding the maturing egg within the developing follicle in the ovary.

cytoplasm: The collection of proteins, carbohydrates, and other molecules and structures that constitute the cell outside of the nucleus.

cytoplast: A cell after removal of its nucleus.

cytostatic factor: The term used to describe the proteins that bring about arrest of the cell cycle; for eggs in metaphase II arrest, the protein cMos is a component of cytostatic factor.

dedifferentiation: The process of losing specialized cell functions and gaining the capacity to divide and give rise to a wide range of alternate cell types.

delayed implantation: A delay of the process of implantation into the lining of the uterus past the normal period of time following fertilization; many mammals have delayed implantation of eggs fertilized during the time they are producing milk for their young; during the delay, the fertilized egg may develop to the blastocyst stage and then enter a period of relative quiescence until lactation ceases, at which time implantation proceeds normally.

dermoid cyst: A term used to describe a fluid-filled, benign tumor of the ovary, or occasionally skin, composed of partially differentiated cells from ectoderm and occasionally endoderm or mesoderm.

diabetes: A disease characterized by the body's inability to properly metabolize blood sugars. Type I diabetes is caused by the loss of insulin-producing cells in the pancreas. Type II diabetes is more common and results from a wider range of cellular disorders.

diacylglycerol: Glycerol with acetyl groups at C1 and C2, an activator of protein kinase C.

differentiation: The typical course of embryogenesis involves the specialization of stem cells into progenitor and precursor cells and finally into adult tissue through the process of differentiation, or specialization.

6-dimethylaminopurine (DMAP): Adenine with two methyl groups at N6, inhibits protein phosphorylation by protein kinases.

diploid: Containing both copies of chromosomes of each pair, generally one from the mother and one from the father.

dizygotic twins: Twins arising from two fertilized eggs; also called fraternal twins.

dopamine: A catecholamine neurotransmitter largely localized to cells in the substantia nigra and striatum. The death of these cells is the cause of Parkinson's disease.

dorsal: A position near the back (top) of an organism.

Down syndrome: A developmental disorder caused by an extra chromosome 21, also called trisomy 21.

dyskinesia: Difficulty in controlling movement; drugs used in the treatment of some diseases like Parkinson's may cause dyskinesia as a side-effect.

electrofusion: Membrane fusion caused by a small jolt of electricity.

embryo: The developmental stage that follows implantation and precedes organogenesis.

embryonal carcinoma: A malignant tumor arising from embryonic cells, usually germ cells.

embryonic diapause: The relatively quiescent period of embryonic development that accompanies an hormonally induced biological delay in implantation; an example is the three-week delay in implantation associated with lactation in a mouse.

embryonic stem cells: Undifferentiated cells derived from the inner cell mass of a blastocyst that will give rise to every cell of the adult organism; in laboratory culture, embryonic stem cells can be maintained in an undifferentiated, dividing state indefinitely.

endogenous retroviruses: Genetic copies of retroviruses that are thought to be due to ancient infections that have become a permanent part of the organism's genome; under certain conditions in the cell, the virus genes may be turned on and produce virus particles.

endometrium: The lining of the uterus that thickens in preparation for pregnancy.

endoreduplication: The replication of chromosomes without cell division.

enucleated: The removal of the nucleus from a cell; the term is also applied to removal of chromosomes in a metaphase plate.

enzyme: A protein that enables chemical reactions to occur at body temperatures.

epiblast: The ectodermal layer of inner cell mass.

epidermal growth factor: One of the first growth factors to be discovered, it acts on a variety of cell types besides epithelial, including neurons.

epididymis: The highly convoluted tubular organ that collects and stores spermatozoa from the seminiferous tubules of the testis.

epigenetic: Specific modifications to gene expression that do not involve alterations in the sequence of bases in the DNA.

ES cells: Embryonic stem cells, typically cultivated from the inner cell mass of a blastocyst.

estrogen: A steroid hormone produced from testosterone by specialized cells in the ovary and some other tissues.

ethics: A system of values and rules of conduct that attempts to discriminate between good and bad behavior.

eugenics: "True" genes, a theory calling for the use of scientific principles to select toward a given set of desired traits.

feeder layer: A layer of cells in the culture dish that is living but has been treated (e.g., exposed to X-rays to damage the DNA) to prevent cell division.

fertilization: The process of sperm entry into an egg followed by formation of individual pronuclei, DNA replication, and cell division to create two new cells with new chromosome combinations.

fetal stem cells: The progenitor cells giving rise to the tissues and organs of a fetus.

fetus: The developmental stage of growth following organogenesis.

FGF4: Fibroblast growth factor number four, discovered to be an important growth factor for stem cells.

fibroblast: Progenitor cells derived from the mesoderm; fibroblast cells derived from mice are used as a feeder layer for stem cells in laboratory culture.

follicle: The structure within the ovary that houses the maturing egg; at the beginning of the monthly menstrual cycle, cells associated with the egg are stimulated to divide and secrete estrogen, which further stimulates their division and the accumulation of fluid, thus creating a blister-like structure about 2 centimeters in diameter at the time of release of the egg at mid-cycle.

follicle stimulating hormone: The protein hormone synthesized and secreted by the pituitary gland of men and women that binds to receptors in both the testis and ovary; classified as a gonadotropin.

forebrain: The prosencephalon, from which develops the telencephalon and the diencephalon, from which develops the cerebrum, the thalamus, and the hypothalamus.

FSH: Follicle stimulating hormone.

galactosidase: An enzyme that hydrolyzes sugars that contain galactose; the gene that encodes it, lacZ, is a bacterial gene that is commonly used as a reporter gene in mammalian cells.

GDF3: Growth differentiation factor 3, a member of the bone morphogenetic protein (BMP) family, more abundant in undifferentiated stem cells.

gene: A specific combination of deoxynucleotides that encodes either messenger RNA, ribosomal RNA, or transfer RNA.

gene pool: A term that refers to the collection of all genes in a given population.

gene replacement: Replacing the gene within the cell with a new gene or partial gene sequence.

genome: Refers to all the genetic information within an organism.

germ cells: Eggs and sperm and their progenitor cells.

germinal vesicle: The nucleus of a growing oocyte.

gestation: The period of embryonic and fetal development within the uterus.

golgi apparatus: The membrane structure responsible for protein processing and transport throughout the cell.

gonadal ridge: The region in the posterior mesoderm that accumulates germ cells migrating from the yolk sac.

gonadotropin releasing hormone: A small protein synthesized and secreted by the hypothalamus that stimulates the pituitary to release gonadotropins.

gonadotropins: Hormones that stimulate the ovary and testis.

gonads: General term for both male and female germ-cell-producing organs and their embryologic precursors.

granulosa cells: Steroid-hormone-producing cells that surround the oocyte and line the growing follicle within the ovary.

growth factors: General term for a large family of proteins or other molecules that stimulate receptors and intracellular pathways that ultimately lead to cell growth and division; may also have other effects on cells.

growth hormone: A small protein secreted by the pituitary that stimulates many tissues and cells by binding to specific receptors, most notably at the ends of long bones in the legs and arms.

growth hormone releasing hormone: A small protein secreted by the hypothalamus that causes release of growth hormone from the pituitary.

gynegenote: The zygote and blastomeres that arise following cleavage of an egg that has been manipulated to contain two female pronuclei.

haploid: Containing only one copy of each chromosome of the species.

hatching: The process of blastocyst escape from the zona pellucida.

hemangioblast: A blood cell progenitor that appears early in embryogenesis.

hemophilia: A disease characterized by absence or malfunction of one of the many factors involved in forming a blood clot; affects mostly men.

hepatic: Pertaining to the liver.

heteroploidy: The state of containing other than the normal number of chromosomes for the cell.

hFES cells: Human embryonic stem cell derived from the inner cell mass of a blastocyst formed following fertilization of an egg by sperm.

hindbrain: The metencephalon and the myelencephalon, from which develop the adult cerebellum, pons, and medulla oblongata.

hippocampus: A curved, elongated ridge of cells located on the floor of each lateral ventricle of the brain, plays a central role in memory.

homolog: Refers to gene sequences or proteins that share a common evolutionary or ancestral origin.

homologous chromosome: Refers to one of the chromosomes of a pair, each of which was derived from a different parent.

homologous recombination: Genetic substitution that results from the binding of specific DNA sequences at the ends of newly introduced DNA that leads to replacement of the original DNA in a chromosome with the new DNA.

hormone: A substance, generally a protein or a steroid, that is synthesized and secreted by body tissues, usually referred to as glands, that can elicit a specific action from nearby cells or distant cells expressing the appropriate receptor.

hPS: Acronym for human stem cells derived from parthenogenetic activation of an unfertilized egg.

human cloning: Refers to the creation of a new human being genetically identical to an existing human by the same methods used to create cloned animals; also sometimes used to describe the creation of laboratory lines of identical human cells.

human immunodeficiency virus (HIV): A retrovirus that is responsible for acquired immunodeficiency syndrome (AIDS); HIV enters cells through receptors meant to elicit immune response, synthesizes a DNA copy of its RNA genome that can integrate and becomes a permanent part of the cell's chromosomes.

hypermethylated: DNA sequences with greater than average numbers of methyl groups added to C5 in cytosine residues to influence expression of the gene.

hyperstimulation: A condition of overresponse to a stimulus such as a hormone.

hypothalamus: An area of the brain beneath the thalamus that carries out several functions to control metabolism including stimulating release of hormones from the pituitary.

ICM: Inner cell mass, the group of cells within the blastocoel of the blastocyst.

immune: Protected against infection by specific responses from the immune system.

immune response: The production of antibodies and other molecules of defense by the blood cells and lymphatic cells that constitute the immune system of an organism.

immune system: Comprising many cell types and structures, including white blood cells and tissue macrophages, most of which are generated by hematopoietic stem cells in the bone marrow.

immunodeficiency: The condition of being incapable of mounting a normal, effective immune response against a pathogen.

implantation: The process of attachment by an early embryo, usually at the blastocyst stage, to the endometrial cells lining the uterus followed by immediate outgrowth of embryonic trophoblast cells between the endometrial cells.

imprinting: The process of influencing gene expression without altering the nucleotide sequence within the gene; termed epigenetic control on gene expression, imprinting occurs differently within the testis and ovary.

inheritance: The process of passing characteristics from parents to progeny.

inositol trisphosphate: A six-carbon sugar with phosphate residues at C1, C4, and C5 that is freely soluble in cytoplasm and interacts with a specific receptor to stimulate release of calcium ions from the endoplasmic reticulum.

institutional review board: A committee of experts that reviews research on human subjects for safety and information value.

insulin: A peptide hormone secreted by the pancreas and used by the body to process sugars. A lack of insulin resulting from defects in pancreatic cells is the cause of type 1 diabetes.

insulin-like growth factor 2: An abundant circulating protein related to insulin that is secreted by adult brain, kidney, pancreas, and muscle.

integration: The process of insertion of new DNA sequences into the DNA strands that comprise the chromosomes.

interphase: The phase of the cell cycle in which the nucleus is visible.

IRB: Internal (or investigative) review board; common acronym for the group of professionals that review research protocols that involve human subjects.

ischemic: The effects of ischemia, a loss of blood supply.

islets of Langerhans: Small groups of cells in the pancreas that produce and release insulin.

IVF: A common acronym for the assisted reproduction technology, in vitro fertilization, which involves recovery of eggs from ovaries and combining them with sperm in laboratory culture to promote fertilization and egg cleavage.

karyoplast: Term used to describe a nucleus that has been removed from the cytoplasm of its cell.

kinase: An enzyme that adds phosphate groups to amino acid side chains in a protein.

kinetochore: A protein motor attached at one end to the centrosome of chromosomes and at the other end to the mitotic spindle.

Krabbe's disease: A rapidly progressive loss of myelin with onset in infancy.

lactate: The production of milk by mammary glands; the ester of lactic acid.

lactational delay: The suspension of embryonic development during the period of milk production.

lacZ: The gene that produces beta-galactosidase.

leukodystrophy: A progressive loss of myelin from the brain, spinal cord, and peripheral nerves.

LH: Luteinizing hormone.

Lou Gehrig's disease: Amyotrophic lateral sclerosis, a disease of the motor tracts of the spinal cord.

luteinizing hormone: A protein hormone synthesized and secreted by the pituitary gland in the brain that stimulates both the ovary and the testis; a gonadotropin.

MAP kinase: The enzyme that adds phosphate groups to mitosis activating protein, which is itself also a kinase.

MAPs: A new term, mature adult progenitor, used to describe stem cells from adult bone marrow with the capacity to differentiate into a wide range of cell types.

maternal messages: Messenger RNAs expressed exclusively, or at abundant levels, in eggs.

maturation promoting factor: The term used to describe the cytoplasmic factors responsible for bringing about the resumption of meiosis in frog oocytes; comprising several cell factors, including CDC/cyclin complexes.

meiosis: The process of dividing chromosomes twice without an intervening cycle of DNA replication and chromosome duplication; results in cells with a haploid number of chromosomes.

mesenchyme: A descriptive term for cells and extracellular connective proteins that constitute the unspecified areas of the developing embryo and fetus.

mesoderm: The middle layer of cells of the three germ layers of a developing embryo. It gives rise to blood, muscle, and skeletal tissue.

metaphase: The stage in the cycle of a dividing cell at which chromosomes have fully condensed and become attached to the proteins that constitute the mitotic spindle but have not begun the process of dividing apart into alternate halves of the cell.

metaphase plate: The imaginary plane in which chromosomes are positioned at metaphase.

micromanipulators: Instruments associated with microscopes that are capable of holding tools, usually glass, and adjusting their position within dimensions smaller than individual cells.

micron: A micrometer; one thousandth of a millimeter, approximately 1/25,000 of an inch.

micropipettes: Hollow tools held by micromanipulators that are capable of delivering or recovering microliter amounts of substances, cell fractions, or cell organelles.

midbrain: The mesencephalon, from which develops the adult aqueduct of Sylvius that connects the ventricles and part of the cerebrum.

mitochondria: The self-replicating, membrane-bound structures within the cell that generate adenosine triphosphate (ATP), which is the principal source of energy for the cell; also contain their own DNA that encodes for some mitochondrial-specific proteins and have been used as an evolutionary tool to prove that most of the mitochondria in an organism derive from the egg, not the sperm.

mitosis: The process of dividing chromosomes, cytoplasm, and cell membranes into two equal daughter cells.

monozygotic twins: Two offspring arising from one fertilized egg.

multiple sclerosis: A chronic disease of the central nervous system in which the loss of patches of myelin interferes with nerve conduction.

multipotent adult progenitor cell: A rare cell in bone marrow that proliferates without aging and maintains pluripotency.

myelin: A specialized, low-conductance cell membrane wrapped around vertebrate axons that serves to insulate and protect nerve impulses; produced by glial-type cells termed oligodendrocytes in the central nervous system and Schwann cells in the peripheral nervous system.

NANOG: A transcription factor that is more abundant in undifferentiated stem cells.

nerve growth factor: A peptide that promotes the growth of nerve tissue. NGF can be used in vitro to encourage the differentiation of nerve cells from embryonic stem cells.

nestin: An intracellular protein expressed by neural cells.

neuron: A nerve cell of the brain or spinal cord that conducts chemoelectric signals.

neurospheres: The ball of cells arising in laboratory culture in the presence of nerve growth factor that consists of neurons, astrocytes, and other glial cells clumped together.

neurotransmitters: The small molecules released by neurons at a gap called a synapse that connects to the target nerve cell; different types of nerve cells secrete different neurotransmitters, including acetylcholine, adrenaline, serotonin, and dopamine.

nidation: The process of communication between the early embryo and the uterine cells, attachment to the uterine cells, invasion of the embryonic trophoblast cells, and the early formation of the placenta.

notochord: The embryonic structure that defines the spinal cord region of the developing fetus.

nuclear membrane: The membrane that separates the nucleus from the cell cytoplasm.

nuclear reprogramming: Altering gene expression to a less differentiated, more pluripotent state, usually by transferring a nucleus from a differentiated cell into an egg.

nuclear transplantation: The transfer of a nucleus from one cell to another; also frequently used to describe the fusion of a somatic cell with an egg, which results in transfer of both the somatic cell nucleus, cytoplasm, and plasma membrane.

nuclear transplantation stem cells: The pluripotent cells arising from blastomeres stimulated to divide following transfer of a somatic cell nucleus, or fusion of the somatic cell, into an egg cytoplast that is then activated to initiate DNA replication and cell division.

nucleolus: The nuclear region for transcription of ribosomal RNAs and assembly of ribosomal subunits.

nucleus: The structure within the cell that contains all the chromosomes and provides the membrane scaffolding required for DNA replication and transcription.

OCT4: A transcription factor that is essential for the development of the inner cell mass.

oligodendrocyte: A type of glial cell in the brain and spinal cord that sends processes out to nerve cells and wraps around and coats the axon in a "myelin sheath," which insulates the axon, permitting low-loss electrical conduction along its entire length.

oocyte: General term to describe all the stages of egg cell growth and maturation.

oogonia: Egg progenitors in the fetus that divide to provide a population of eggs for the ovary.

organ: A collection of specialized cells and tissues that carry out a defined function within the body, such as the heart.

organ transplantation: The surgical insertion of a new organ to replace, or assist, an existing organ; new kidneys are frequently inserted into the blood supply of an existing kidney without removing the defective kidney.

organism: Any living thing, plant or animal, single or multicelled.

ovarian follicle: The blisterlike structure within the ovary that enlarges in response to the hormone signals that stimulate cell division by the granulosa cells that surround and support the final stages of maturation by an egg.

ovasome: A new term coined to describe the use of eggs to create somatic cells for stem cell therapy rather than new offspring.

ovulation: The process of final maturation and release of an egg from the ovary.

pancreas: The organ that secretes insulin in response to blood sugar.

Parkinson's disease: A disease of the loss of function of dopamine-producing nerve cells located in the midbrain that results in limb tremor and slow, stiff motion.

parthenote: The blastomeres and cells that arise following egg activation in the absence of sperm.

pathogen: Any microorganism, such as a bacterium or virus, that causes disease.

perinatal: Pertaining to the few weeks of time before and after birth.

perivascular spaces: Areas in the brain immediately surrounding blood vessels.

perivitelline space: The space between the plasma membrane of the egg and the inner aspect of the zona pellucida that surrounds the egg.

phosphatase: An enzyme that removes phosphate groups from proteins.

phospholipase C: One of a family of enzymes that cleaves bonds in phospholipids to generate two second messengers, diacylglycerol and inositol trisphosphate.

pituitary: The gland in the brain near the hypothalamus that synthesizes and secretes several protein hormones, including growth hormone and the gonadotropins, follicle stimulating hormone, and luteinizing hormone.

plasma membrane: The membrane that encloses and separates the cell from the outside environment.

pluripotent: Cells that have the ability to differentiate into more than one type of tissue—like embryonic stems cells—are pluripotent. ES cells, because they can't form trophoblast tissue, are not considered to be totipotent, or capable of producing an entire organism.

polar body: The small cell that results from the asymmetric division of the egg; the first polar body contains the diploid set of chromosomes resulting from the first meiotic division; the second polar body contains the haploid set of chromosomes resulting from the second meiotic division.

polar trophectoderm: The layer of outside cells that overlies the inner cell mass cells in the blastocyst.

polyadenylation: The protein synthesis machinery within the cell depends on a series of adenosine molecules at the end of a messenger RNA to stimulate translation of the message into protein; although there are exceptions, in general, messenger RNAs that are not polyadenylated are in a state of storage and are not expressed as protein.

polycystin-1: A large cell-surface glycoprotein with repeating units that determine its three-dimensional structure.

polygamy: The social practice of marrying more than one person.

polyploid: A cell containing more than one pair of haploid chromosomes.

precursor: The cell immediately preceding the final stage of differentiation.

precursor cell: A cell, typically derived from a progenitor cell, that is fated to differentiate into one or two specific cell types.

pre-implantation: The period of development of an organism that precedes implantation in the uterine lining.

primary oocyte: The stage of oocyte maturation that includes a not fully grown egg arrested in the prophase of meiosis and surrounded by one layer of granulosa cells.

primitive ectoderm: The layer of inner cell mass cells immediately under the polar trophectoderm of a newly implanted blastocyst that is forming a bilaminar embryo.

primitive endoderm: The layer of cells that lines the inner cell mass and the blastocoel cavity on the inner aspect of the blastocyst.

primitive node: The pit that develops on the dorsal side of the bilaminar embryo as a result of outgrowth and expansion of mesodermal cells between the primitive ectoderm and the primitive endoderm resulting in a trilaminar embryo.

primitive streak: The groove that develops from the primitive node to the outer edge of the embryo at the transition from a bilaminar to a trilaminar stage of development; the appearance of the primitive streak defines the embryo with respect to dorsal, ventral, caudal, and cephalic regions.

progenitor: An ancestral cell committed to the direct line of cell differentiation.

progenitor cell: Multipotent cells that can differentiate into a limited set of cell types.

progressive multifocal leukoencephalopathy: loss of myelin in the central nervous system caused by a viral infection.

pronucleus: The large nucleus with a few, prominent nucleoli, formed by an activated egg to contain either egg chromosomes or the sperm head, if fertilized, or formed by remodeling a somatic cell nucleus transplanted into the egg.

prophase of meiosis: The period between replication of the complete set of chromosomes and the onset of meiosis during which chromosomal crossing-over occurs; in most mammals, including humans; this period extends from fetal development to puberty.

protamine: A relatively low molecular weight protein, rich in arginine.

puberty: The onset of sperm production in males and oocyte maturation and release in females.

quickening: The time during a pregnancy when the fetus has grown to a size sufficient for the mother to feel its movements within the uterus.

reporter gene: A well-characterized, easily detectable gene that is attached to a gene of interest; because of their close association, if the reporter gene is expressed, the gene under study is also assumed to be expressed.

retinal degeneration: A disease process that results in loss of retinal cells in the eye and consequent loss of vision.

retinoic acid: A small molecule related to vitamin A that plays a major role in guiding development of early embryonic structures.

retroelement: A general term for retrovirus-related sequences in the chromosomes of all mammalian cells.

retrovirus: A virus whose genome is made of RNA and replicates by causing the synthesis of a DNA copy of its RNA genome that then serves as template for new RNA synthesis for progeny virus.

ribosomes: The two-subunit, protein-RNA structure responsible for organizing all the intracellular elements necessary to compile the specific amino acid sequences of proteins as directed by the nucleic acid sequences in the messenger RNA.

"S" phase: The DNA synthesis phase of a cell cycle.

second messenger: A term applied to a wide variety of intracellular molecules that promote an action by the cell in response to binding of extracellular molecules, such as hormones or growth factors, to receptors in the plasma membrane.

seminomas: Cancers of the testis arising from sperm progenitor cells.

somatic cell: A body cell, as opposed to a reproductive (germ) cell like an egg or a sperm.

sperm: The male germ cell.

spermatogenesis: The process of creating new sperm within the seminiferous tubules of the testis.

spermatogonia: The progenitor sperm cell that continually divides to produce new sperm.

spermidine: A polyamine with various cellular functions, originally identified in semen.

spina bifida: A developmental defect in which the spinal column fails to close completely leading to protrusion of the spinal cord or the membranes that surround it.

STELLAR: A DNA binding protein, 159 amino acids, related to mouse STELLA that is expressed in developing germ cells.

stem cell: An uncommitted cell that can divide continually without loss or gain of chromosomes or genes and gives rise to every cell type in the body.

stroke: An acute loss of brain function resulting from blockage or rupture of a blood vessel.

subgranular zone of the hippocampus: A layer of neuronal progenitor cells within the dentate gyrus, a region of the hippocampus.

subventricular zone: A thin layer of dividing, neural progenitor cells along the lateral wall of the lateral ventricles of the brain that give rise to neurons destined for the olfactory bulb.

syngeneic: Arising from the same or genetically identical organism.

syrinx: A tube-shaped, fluid-filled swelling in the spinal cord that accompanies spinal injury; may prevent nerve regeneration at the site and may produce pain and weakness.

teratoma: Tumor, usually benign, with partially differentiated cells from all three embryonic cell layers, ectoderm, mesoderm and endoderm.

tetraploid: Four copies of each chromosome.

therapeutics: A substance used to correct a disease state.

tissue: A collection of cells coordinated to perform a specific body task.

tissue transplantation: The process of transferring cells, such as bone marrow cells, from one person to another to correct a disease in the recipient.

transcription: The process of creating an RNA copy of a DNA sequence in a gene.

transdifferentiation: Changing a cell, or its daughter cells, from one differentiation state to another.

transfection: A process that inserts new DNA into a cell.

transgenic: An organism with foreign DNA in its cells that was introduced at or near the time of fertilization.

transverse myelitis: An acute disease of both sides of the spinal cord with an unknown cause.

trimeric G protein: A three-subunit protein linked to a huge class of cell surface receptors, G protein-coupled receptors, that bind a wide class of signaling agents from odors to hormones.

triplet codon: The group of three nucleotides in a messenger RNA that determines which amino acid to add to the growing protein within the ribosomal complex; more than one codon can define each amino acid, and some codons signal the beginning and ending of the protein.

trophoblast: The outside cells of a blastocyst that will give rise to the placenta and contribute to other extra-embryonic membranes such as the amnion.

ultrasound waves: Sound waves with a wavelength too short to be heard by the human ear.

ventral: A position near the belly of the body or the anterior of an object.

von Wildebrand's disease: A disease of the blood clotting system that involves defects in platelet aggregation and affects both men and women.

zona pellucida: The glycoprotein coat synthesized by the egg that surrounds it and protects it during the pre-implantation period of development.

Index

of mice, 114
    nuclear transplant and, 9–10, 58, 153–157
    from somatic cells, 158–159
    success rates of, 157, 161
clotting factor, 226–227
cMos, 45–47
    degradation of, 60–61
    egg activation and, 53–54
    in metaphase II, 52
COH. *See* controlled ovarian hyperstimulation
    (COH)
commitment events
    first, 120–121
    in stem cell differentiation, 188–192
compaction, 123
constitutive expression, 110
controlled ovarian hyperstimulation (COH),
    42–43
    risks of, 43–44
Cooper, Geoffrey, 46
cord blood stem cells, 172–173, 175–176
corona radiata, 41
corpus luteum, 137
Crick, Francis, 17
cross-species contamination, 210, 216–217
CSF. *See* cytostatic factor (CSF)
culture media, 94–97, 96
cumulus cells, 41
cyclic adenosine monophosphate (cAMP), 51
cyclic AMP, 44
cyclin B, 44–47
cyclin-dependent kinases (Cdks), 26, 28
cyclin genes, 27
cyclins, 26, 28
Cyno-1 stem cell line, 131
cytoplasts, 58, 153
cytosine (C), 22, 86
cytostatic factor (CSF), 45

Daley, George, 221
Das, Gopal, 196–197
daughter cells
    asymmetric cell division of, 189–190
    in bone marrow stem cells, 170

    clonal expansion of, 171
    definition of, 16
    in pronuclear eggs, 67
Davis, Gray, 239
DDK mouse strain, 69
death, 244
dedifferentiation, 102, 104
delayed implantation, 128–129
dementia, 204–205
dermoid cysts, 52–53
diabetes, 217
    type I, 217–220
    type II, 217, 220–221
diacylglycerol, 59
Dickey-Wicker Amendment (1996), 237–238
differentiation
    back, 121
    cell function genes and, 110–111
    definition of, 5–6, 21
    in embryo development, 144
    into neurons, 201–203
    nuclear remodeling and, 110–114
    nuclear transplantation and, 102–104,
        153–157
    prevention of, 176–177
    in stem cells, 188–192
    transdifferentiation, 171
diseases
    as cellular problems, 2–3
    neurogenerative, 196–214
    tissue systems failures, 216–230
dissecting microscopes, 102
DMAP. *See* [six]6-dimethylaminopurine
    (DMAP)
DNA (deoxyribonucleic acid)
    blastomere replication of, 98
    discovery of, 17
    endoreduplication, 124
    error detection checkpoints, 66
    methylation of, 69
    mitochondrial, in tissue rejection, 223
    origin of replication, 19
    replication, 98
    replication controls on, 18–19